ADVANCES IN
SONOCHEMISTRY

Volume 5 • 1999

ADVANCES IN SONOCHEMISTRY

Editor: TIMOTHY J. MASON
School of Natural and
Environmental Sciences
Coventry University
Coventry, England

VOLUME 5 • 1999

JAI PRESS INC.
Stamford, Connecticut

Coventry University

CONTENTS

LIST OF CONTRIBUTORS

Geoffrey E. Attenburrow   British School of Leather Technology
                             Nene College of Higher Education
                             Northampton, England

Pedro Cintas   Departamento de Química Orgánica
                             Universidad de Extremadura
                             Badajoz, Spain

Ji-Feng Ding   British School of Leather Technology
                             Nene College of Higher Education
                             Northampton, England

Thierry Lepoint   Laboratoire de Sonochimie
                             d'Etude de la Cavitation Institut Meurice
                             Brussels, Belgium

Françoise Lepoint-Mullie   Laboratoire de Sonochimie
                             d'Etude de la Cavitation Institut Meurice
                             Brussels, Belgium

Jean-Louis Luche   Laboratoire de Chimie Moléculaire et
                             Environement
                             ESIGEC-Université de Savoie
                             Le Bourget du Lac, France

Gertraud Mark   Max-Planck Institute for Radiation Chemistry
                             Mulheim an der Ruhr, Germany

Timothy J. Mason   School of Natural and Environmental Sciences
                             Coventry University
                             Coventry, England

Sukhvinder S. Phull   School of Natural and Environmental Sciences
                             Coventry University
                             Coventry, England

John P. Russell   ARC Sonics Inc.
                                        Vancouver, British Columbia
                                        Canada

Heinz-Peter Schuchmann   Max-Planck Institute for Radiation Chemistry
                                        Mulheim an der Ruhr, Germany

Martin Smith   Chipping Campden
                                        Gloucestershire, England

Clemens von Sonntag   Max-Planck Institute for Radiation Chemistry
                                        Mulheim an der Ruhr, Germany

Armin Tauber   Max-Planck Institute for Radiation Chemistry
                                        Mulheim an der Ruhr, Germany

Maricela Toma   Costin D. Ninitzescu Institute of Organic
                                        Chemistry
                                        Bucharest, Romania

Mircea Vinatoru   Costin D. Ninitzescu Institute of Organic
                                        Chemistry
                                        Bucharest, Romania

Jian-Ping Xie   British School of Leather Technology
                                        Nene College of Higher Education
                                        Northampton, England

PREFACE

This is the fifth volume of *Advances in Sonochemistry*, the first having been published in 1990. I must confess that in compiling the contributions for the current volume I have been influenced by the way in which my own definition of sonochemistry has changed to include not only the ways in which ultrasound has been harnessed to effect chemistry but also its uses in material processing. Thus the reader will find that within these seven chapters are included three which might be considered to be in the main stream of "chemistry" (sonoluminescence, chemical dosimetry, and organic synthesis) and one chapter represents a borderline between chemistry and processing (ultrasound in microbiology). Of the remainder two are in the domain of processing (ultrasound in the extraction of plant materials and in leather technology) and a final chapter which describes a new type of large scale equipment which uses audible frequencies in processing applications. As with previous volumes, the contributors come from a number of countries: Belgium, Canada, France, Germany, Romania, Spain, and the U.K. The geographical spread also reflects the continuing major interest of European laboratories in sonochemistry.

Sonoluminescence has been an interest for both chemists and physicists for many years. In recent times, however, interest has grown as instrumentation has permitted greater probing of the phenomenon and mathematicians have brought in their own interest and expertise in bubble dynamics. This has led to rather advanced treatments of sonoluminescence at conferences and in scientific contributions. With this in mind I thought that it would be of use to those sonochemists whose main interests are not in this field to include a general introduction to the subject. Any such introduction will naturally carry with it the views of the writers, but I believe that Thierry Lepoint and his wife Françoise Lepoint-Mullie have done an outstanding job in bringing together so much background information covering nearly 200 references.

Dosimetry continues to interest all sonochemists in that we all would like to have a method of quantifying the effects of sonication. A range of methods are available which can be approximately divided into physical methods (e.g. calorimetry and acoustic pressure measurements) and chemical methods (e.g. iodimetry and hydroxy radical trapping). When sonochemistry is used to promote chemical effects such as free radical polymerization and advanced oxidation for water treatment the chemical dosimeter is clearly important. Radiation chemists have been involved in OH radical formation and measurements for many years, and so it was important to obtain a contribution from Clemens von Sonntag who is an expert in this area and has recently turned his attention to dosimetry in the sonolysis of aqueous solutions.

The title of the third contribution "Can Sonication Modify the Regio- and Stereoselectivities of Organic Reaction?" is self explanatory. Jean-Louis Luche and Pedro Cintas have attempted to answer the question but at this stage in the development of sonochemistry it is not possible to be absolute in the answer. Nevertheless the concept is important to organic synthesis because it is clear that sonication can influence selectivity. Hence a review of current literature on the subject is important.

As far as I am aware there has been no general review published on the uses of ultrasound in microbiology. There are many topics which have been gathered together in this chapter each of which are of current interest including water remediation, biotechnology, and medicine. In writing this chapter my colleague, Suki Phull, and I were aware that the readership would mainly be chemists rather than biologists. For this reason we have included a lengthy introduction to microbiology and some general definitions to help the reader. I am confident that this will be a major growth area for sonochemical studies in the future.

With the interest in natural rather than synthetic remedies for ailments there is an increasing attention paid to plant products. Coupled with this is the requirement that better techniques should be developed in order to maximize extraction of the target materials. With the help of my colleagues Mircea Vinatoru and Maricela Toma from Romania we have prepared a contribution on the ultrasonically assisted extraction of bioactive principles from plants. Romania is a country where plant extracts have been used for generations and so there is a wealth of background information on plant varieties which are useful in terms of their extracts. The scale-up of ultrasonic extraction has been achieved and so the positive laboratory results included in this chapter should soon be realized industrially.

Most chemists have only a sketchy knowledge of what is involved in leather processing. This is perhaps not unexpected since it is not a topic which appears in standard undergraduate chemistry, or materials, courses. In taking a raw hide through to the finished leather product most of the processes involve leather being immersed in a solution of chemicals and so it is reasonable to suppose that sonication should have an effect in terms of both improved penetration and surface reactions. A few years ago I became involved with some collaborative research involving the British School of Leather Technology and, as a result of this, I asked my colleagues Ji-Feng Ding and Geoff Attenburrow to contribute a chapter on the uses of power ultrasound in leather technology. I hope that readers will see that while the topic addresses leather processing the potential applications spread much more widely than this.

The final contribution is somewhat different in style from those which precede it. I asked John P. Russell and Martin Smith to describe the processing equipment which has been developed by Arc Sonics in Canada. There were two reasons for this. First, the operation of the equipment is based on acoustic resonance but at audible rather than ultrasonic frequencies, and second that it has been operated on a large, industrial scale. A number of its applications seem to mirror those of ultrasonic sonochemistry and the basis of its action appears to be cavitation. Because of the novelty of its design the chapter details its mode of operation and potential rather more than the fundamental chemistry involved. Over the next few years it is likely that there will be many more applications for sonochemistry on a large scale and at low frequency.

Timothy J. Mason
Series Editor

AN INTRODUCTION TO SONOLUMINESCENCE

Thierry Lepoint and Françoise Lepoint-Mullie

OUTLINE

Advances in Sonochemistry
Volume 5, pages 1–108.
ISBN: 0-7623-0331-X

ABSTRACT

Sonoluminescence—the emission of light associated with the collapse of cavitation bubbles in a state of oscillation which are forced by an acoustic pressure field—occurs in the case of multibubble clouds and single bubbles maintained in levitation in a liquid. In this chapter, we focus on single-bubble sonoluminescence. After a short introduction (Section 1), the behavior of an isolated bubble driven by an acoustic wave is reported on, so that the main mechanical events invoked as triggering off sonoluminescence are introduced straight away (Section 2). These mechanisms—applied to different species (i.e., molecules in the interior gas and/or liquid phase near the bubble–liquid interface) and for which there is no consensus at the present time—are presented as a function of their increasing complexity, i.e., (1) spherical collapse with uniform intracavity pressure and emission of a strong pressure pulse in the liquid, (2) as just stated, but with nonuniform intracavity pressure (the formation of an inward shock or pressure wave), and (3) aspherical collapse with the formation of an inward jet that strikes the opposite bubble wall. This presentation, illustrating the great complexity in bubble behavior, is used as a basis for the interpretation of the experimental data reported in Section 3. This enables two main questions to emerge. (1) Are sonochemistry and sonoluminescence correlated? (Section 4). (2) What is the site of sonoluminescent activity (the gas phase , the liquid shell just outside the bubble in collapse, or both)? Although an experimental answer is provided for question (1), it appears from an examination of the literature that the second question is not explicitly answered, so that at the present time the type of emitting species has not yet been identified (Section 5). It also appears that question (2) could be applied to multibubble sonoluminescence.

1. INTRODUCTION

In 1933, Marinesco and Trillat [1] immersed photographic plates in a water bath in which ultrasonic waves were propagated. They noted a blackening of the plates and attributed this effect to oxidoreduction processes stimulated by ultrasound. One year later, repeating the same experiment, Frenzel and Schultes [2] concluded that the darkening of the plates was caused by luminescence from the ultrasonic field. By itself, ultrasound has no direct action on chemical bonds involved in the molecules of the liquid (or the dissolved gas) through which it travels. Indeed in such kinds of experiments the energy density of the sound field is only 10^{-10} eV per atom (1 eV = 1.602×10^{-19} J). In fact, luminescence results from cavitation, i.e., a process in which a great number of bubbles oscillate nonlinearly when produced in a liquid as the acoustic wave propagates. After growth forced by the acoustic field the bubbles collapse violently. This step enables the low energy density of the sound field to be mechanically focused, so that photons with an energy of several electron-volts are emitted. The concentration factor of energy is therefore about 11 orders of magnitude. Photons so emitted are detectable with the naked (but adapted to the dark) eye. In 1939, Harvey [3] suggested using the term *sonoluminescence* (SL) for this weak light triggered at the end of a bubble collapse. At the present time this definition is no longer exclusively associated with acoustic cavitation, but is also connected with hydrodynamic [3–7] or laser- and spark-induced cavitation [8–15].

In this paper we deal with acoustically induced SL which can occur in two situations as shown in Figure 1.

(1) Acoustic waves traveling in a given liquid can be intense enough for numerous gas nuclei to be generated. After this nucleation process, the bubbles can grow, oscillate nonlinearly, and finally collapse. This dynamic behavior strongly depends on the properties of the liquid (surface tension, viscosity, density), the acoustic field (intensity and frequency, continuous or pulsed waves), and the bubbles themselves (gas/vapor content and initial radius) [16]. This situation involving a multibubble field leads to multibubble sonoluminescence (MBSL) and is known for alternate pressure waves with a frequency range of between a few hertz [17] and about 2 MHz. It seems that the frequency range higher than 2 MHz has not been investigated, at least as far as the present authors are aware. Other effects are acoustic emissions (the analysis of which reveals that acoustic cavitation is characterized by deterministic chaos [18]) and sonochemical effects (e.g., the production of radicals, erosion, emulsification, dispersion and the breaking of polymer chains) [19,20].

When bubbles oscillate more than several thousand acoustic cycles, what is involved is "stable cavitation." "Transient cavitation" refers to bubbles, the usual lifetime of which is no more than a few acoustic cycles. Transient cavitation generally occurs for acoustic intensities of 10^4 W/m^2, i.e. acoustic pressures higher than 2 bars (see the equation in note 2).

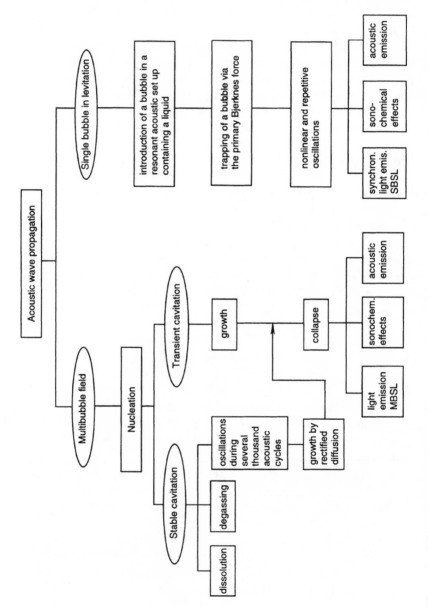

Figure 1. Schematic representation of the two modes of sonoluminescence and their relations with bubble dynamics.

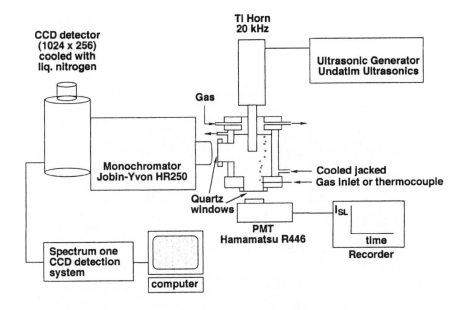

Figure 2. Typical setup for the generation of multibubble SL at low acoustic frequencies (via an immersed titanium horn) (acoustic frequency ~20 kHz).

(2) In 1990, Gaitan and Crum [21] (see also [22]) showed in a remarkable experiment that a single bubble maintained for hours in levitation in a resonant acoustic setup kept emitting extremely short bursts of light. Since then this phenomenon has been called *single-bubble sonoluminescence* (SBSL) and is known for an acoustic frequency range between 20 and 50 kHz and acoustic pressures of between about 1 and 1.5 bars. Both MBSL and SBSL occur with various liquids (water, aqueous solutions, and organic media) and dissolved gases.

MBSL was recently reviewed with special attention to the physical [16,23,24] or physicochemical aspects [25–27]. Taking into consideration that SBSL affords extremely accurate and determining data which cannot be obtained via MBSL studies, we focused particularly on SBSL in this introduction. The reader interested in an advanced overview of SBSL should consult a recent work by Barber et al. [28].

2. BUBBLE HISTORY

2.1 Preliminary Comment

SL is intimately associated with acoustic cavitation. In the same way as photochemists are compelled to follow the laws of quantum mechanics, sonochemists

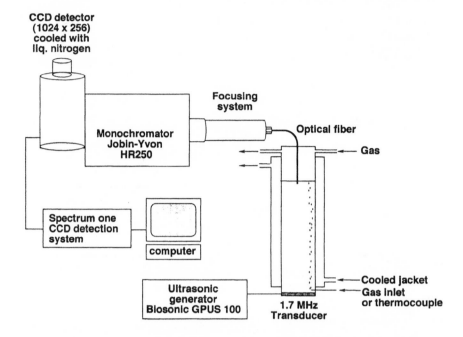

Figure 3. As in Figure 2, but for a high acoustic frequency system (acoustic frequency >100 kHz). The PZT transducers are in immediate contact with the solution.

and sonophysicists are confronted with the laws of fluid mechanics and, in particular, the properties of ultrasonic waves and the dynamics of bubbles, both of which are the source of sonochemical and sonoluminescent activity.

In spite of the possibly arid character that the reading of such an approach may imply, we have chosen to begin with a description of some of the main events that a bubble undergoes, and the related equations. In this way, the extreme versatility of the bubble behavior will emerge. Moreover the presentation of the results, their interpretation and/or extrapolation will be facilitated. However, the busy reader anxious to deal with experimental data can refer to Section 2.9 (and the following) and approach the interpretation of the data by progressively accessing the basic equations and phenomena reported in Sections 2.3 to 2.8.

The reader can refer to an introductory chapter on acoustic cavitation [29] that the present authors recently referred to as a contribution to cavitation free of any ponderous mathematical calculations. Advanced descriptions are found in the works quoted [30] which are considered by the specialists in the field to be authoritative references. The nucleation step which is a characteristic of the generation of multibubble clouds (Figure 1) is not considered in the present text (the reader should consult [29] for an introduction and the work by Atchley and Crum

[30b] and Atchley and Prosperetti [31] for advanced descriptions). Indeed in single-bubble sonoluminescence and sonochemistry, a bubble is introduced "externally," i.e., either via a syringe or a heated Ni-Cr wire. Consequently the sound wave is not used to generate cavitation.

2.2 The Isolated Spherical Bubble: Historical Aspects

Developments in the study of the dynamics of bubbles in liquids date back to 1894, the year of the construction of HMS *Daring*, a destroyer subjected to anomalous vibration and excessive slippage of the propeller that resulted in low speeds [32]. Considering that bubbles created hydrodynamically by the rotation of the propeller subjected it to vigorous erosion, Lord Rayleigh [33] provided a theoretical description according to which bubble collapse creates enormous pressures in the liquid just outside a bubble in collapse, with these pressures acting as "hammers." In fact, the phenomenon of erosion resulting from the presence of bubbles in the vicinity of solids received a new interpretation through the work by Kornfeld and Suvorov [34], who suggested that the asymmetric implosion of bubbles led to the formation of a high-velocity intracavity jet likely to induce erosion. However, using high-speed cinematography, Philipp and Lauterborn [35] recently demonstrated that this jet is not directly responsible for the erosion. In fact, when the tip of the jet joins the opposite face of the bubble, a toroidal bubble forms which, via implosion, generates local microshock waves that induce erosion.

However, Lord Rayleigh's approach constitutes the fundamental basis for the approximate radial motion of spherical bubbles and, consequently, stimulated a great number of refinements developed in order to predict as accurately as possible the behavior of the bubble interior and the liquid shell surrounding the bubble–liquid interface.

2.3 The Rayleigh Collapse and the Rayleigh–Plesset Equation

Simplifying the problem of bubble dynamics by considering an empty spherical isolated bubble (the problem put forward by Besant [36]), Lord Rayleigh [33] set out to determine (1) the duration of collapse, (2) the speed of the bubble wall, and (3) the pressure generated in the liquid near the bubble as a result of an implosion. He considered an empty bubble that he assumed to remain spherical all the time and an incompressible, nonviscous liquid (of specific mass ρ) with a negligible surface tension (σ). The first two questions can be solved as follows.

We consider a bubble of radius R_m as shown in Figure 4. If the bubble was filled with gas and vapor from the liquid, its internal pressure (P_{int}) would be

$$P_{int} = P_{gas} + P_{vap} \tag{1}$$

where P_{gas} and P_{vap} are the gas and the vapor pressure, respectively. Because of the Laplace pressure ($P_{Lap} = 2\sigma/R_m$) the pressure in a bubble at rest is greater than the

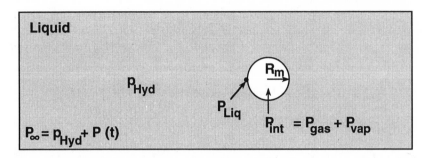

Figure 4. Representation of a spherical bubble of radius R_m in a liquid of viscosity η and surface tension σ. The hydrostatic pressure is denoted p_{Hyd}; P_{Liq} is the pressure in the liquid just outside the bubble; P_{gas} and P_{vap} are respectively the gas and vapor pressure inside it; $P(t)$ is a time-dependent pressure; and p_∞ represents the pressure in any zone of the liquid at a distance from the bubble.

pressure in the liquid immediately adjacent to it. If the pressure just outside the bubble wall is denoted as P_{Liq}, equilibrium is achieved for

$$P_{int} = P_{gas} + P_{vap} = P_{Liq} + \frac{2\sigma}{R_m} \qquad (2)$$

Of course, if the bubble is empty ($P_{int} = 0$) and the surface tension is negligible, P_{Liq} will be zero. It is assumed that at initial time t, the cavity radius is R_m and the bubble wall velocity is zero.

Hydrostatic pressure (p_{Hyd}) exerts a mechanical work which causes the cavity to pass from radius R_m to R. This work is $(4\pi/3)[R_m^3 - R^3]\, p_{Hyd}$ and is found in the form of kinetic energy. This energy is released to several liquid shells of thickness Δr, mass $4\pi r^2 \Delta r \rho$, and the liquid thus acquires speed $\dot{r}$ ($\dot{r} = dr/dt$). Therefore,

$$\frac{4\pi}{3}[R_m^3 - R^3]\, p_{Hyd} = \frac{1}{2}\int_R^\infty \dot{r}^2 \rho 4\pi r^2 dr \qquad (3)$$

Because the liquid is incompressible, the rate of liquid mass flowing through any spherical surface of radius r equicentric with the cavity must be constant. During period Δt, a mass of liquid equal to $(4\pi r^2 \dot{r} \Delta t)\rho$ flows out of the bubble across a surface of radius r. By equating this flow to the one at the surface of the bubble (of radius R), $\dot{r}/\dot{R} = R^2/r^2$ is obtained. After substitution, the integration of Eq. (3) gives the kinetic energy as $E_{kin} = 2\pi\rho\dot{R}^2 R^3$. Since this energy is equal to the work causing the cavity to pass from radius R_m to R, the speed of the bubble wall can be obtained as

$$\dot{R} = \left[\frac{2p_{Hyd}}{3\rho} \left(\frac{R_m^3}{R^3} - 1 \right) \right]^{1/2} \tag{4}$$

The time for a complete collapse (i.e., for the bubble radius to pass from R_m to $R = 0$) is inversely proportional to the bubble wall speed ($\dot{R}$) and proportional to R_m:

$$t_{col} = \int_{R_m}^{R=0} \frac{dR}{\dot{R}} = 0.915 R_m \left(\frac{\rho}{p_{Hyd}} \right)^{1/2} \tag{5}$$

The answer to the third question can be approached as follows. An empty bubble with radius R_m is at rest in a simplified liquid (Figure 4). The hydrostatic pressure is maintained constant but a time-dependent pressure $P(t)$ is superimposed on p_{Hyd}. Thus, the pressure in the liquid at any point at a distance from the bubble is $p_\infty = p_{Hyd} + P(t)$. The bubble undergoes a modification in volume under the effect of this variation in pressure. The liquid shell so involved acquires kinetic energy, the expression of which is determined above (the case of an incompressible liquid). The work which led to a modification in volume is the difference of work applied remote from the bubble by means of p_∞ and the work applied by pressure P_{Liq} at the bubble interface (in the liquid). Therefore,

$$2\pi R^3 \rho \dot{R}^2 = \int_{R_m}^{R} (P_{Liq} - P_\infty) 4\pi R^2 dR \tag{6}$$

This equation can be derived with respect to R. Bearing in mind that by definition, $\dot{R} = dR/dt$, the expression for $d(\dot{R}^2)/dR$ can be easily obtained. Indeed,

$$\frac{d(\dot{R}^2)}{dR} = \frac{1}{\dot{R}} \frac{d}{dt} \left[\frac{dR}{dt} \frac{dR}{dt} \right] = \frac{1}{\dot{R}} \left[\frac{dR}{dt} \frac{d^2R}{dt^2} + \frac{dR}{dt} \frac{d^2R}{dt^2} \right] = 2 \frac{d^2R}{dt^2} = 2\ddot{R}$$

where $\ddot{R}$ is the acceleration of the bubble wall. Consequently, the Rayleigh–Plesset (RP) equation is obtained (Plesset [37] reported an analysis similar to Lord Rayleigh's but involving a vapor bubble):

$$\ddot{R}R + \frac{3}{2}\dot{R} = \frac{1}{\rho}(P_{Liq} - P_\infty) \tag{7}$$

Though the analysis reported here enabled Lord Rayleigh to substantiate his hypothesis that local overpressures in a liquid caused erosion (Figure 5), it appears from Eq. (4) that the bubble wall speed tends to be infinite as the bubble radius tends to zero. Lord Rayleigh suggested that the gas in an actual bubble acted as a cushion limiting the collapse. However, this approach remains confined to the case

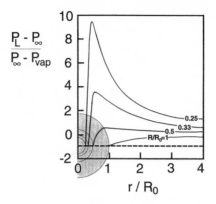

Figure 5. Adimensional representation of the evolution of the pressure in the liquid just outside an empty bubble (P_{Liq}) (the Rayleigh model). The pressure parameters are defined in the caption to Figure 4. The distance from the bubble is denoted r; R_0 is the initial radius; R is the bubble radius at a given time. The detailed equation [similar to Eq. (14), see below] is given in [38, p. 68]. The result for a gas bubble in compressible water is mentioned in Section 5.5.

of a freely oscillating bubble; this is not really the case for a gaseous bubble in an ultrasonic field.

2.4 The Isolated Gas Bubble in Forced Oscillation

Noltingk and Neppiras [39] modified the RP equation [Eq. (7)] in order to describe the radial behavior of a spherical gas bubble driven by an acoustic wave of frequency ν_{ac} traveling over a liquid of surface tension σ.[1] They assumed that the radial motion would be rapid enough for heat exchanges with the surrounding liquid to be negligible (the condition of adiabaticity).

We first mention the role of the intracavity gas in order to calculate P_{Liq} and substitute its expression in Eq. (7). As a second step, we introduce the role of an alternate pressure wave in the term P_∞ of the same equation.

If a "bubble/liquid" system in a state of equilibrium (Figure 4) is considered, it will be noticed that P_{Liq} (the pressure just outside the bubble wall) $= P_{Hyd}$ and P_∞ (the pressure at a very large distance from the bubble) $= P_{Hyd}$. Therefore, in the case of a bubble (of initial radius R_{init}) containing both gas and vapor, we can write

$$P_{gas} + P_{vap} = P_{Liq} + \frac{2\sigma}{R_{init}} = P_{Hyd} + \frac{2\sigma}{R_{init}} \tag{8}$$

so that the gas pressure at initial time is

$$P_{gas} = P_{Hyd} + \frac{2\sigma}{R_{init}} - P_{vap} \tag{9}$$

We consider that the bubble volume changes and the transformation is adiabatic. Therefore, Eqs. (10) and (11) hold where γ is the adiabatic index ($= C_p/C_v$ with C_p and C_v the heat capacities of the intracavity gas at constant pressure and volume, respectively):

$$P_{init}\left[\frac{4\pi R_{init}^3}{3}\right]^\gamma = P_{fin}\left[\frac{4\pi R_{fin}^3}{3}\right]^\gamma \tag{10}$$

or

$$P_{fin} = P_{init}\left[\frac{R_{init}}{R_{fin}}\right]^{3\gamma} \tag{11}$$

Subscripts "init" and "fin" denote the initial and final states, respectively. In these equations P_{init} can be subtituted for P_{gas} [as expressed in Eq. (9)]. The gas pressure after bubble volume change can be expressed as

$$P_{fin} = \left(P_{Hyd} + \frac{2\sigma}{R_{init}} - P_{vap}\right)\left(\frac{R_{init}}{R_{fin}}\right)^{3\gamma} \tag{12}$$

Total pressure (P_{tot}) includes the role of vapor pressure and is found as

$$P_{tot} = \left(P_{Hyd} + \frac{2\sigma}{R_{init}} - P_{vap}\right)\left(\frac{R_{init}}{R_{fin}}\right)^{3\gamma} + P_{vap} \tag{13}$$

Taking into consideration Eq. (2), the pressure in the liquid immediately beyond the bubble interface is

$$P_{Liq} = \left(P_{Hyd} + \frac{2\sigma}{R_{init}} - P_{vap}\right)\left(\frac{R_{init}}{R_{fin}}\right)^{3\gamma} + P_{vap} - \frac{2\sigma}{R} \tag{14}$$

This equation for P_{Liq} can be introduced into the Rayleigh–Plesset equation [Eq. (7)].

The second problem is to represent the effect of an alternate acoustic pressure. In such a case, P_∞ can be represented as follows:

$$P_\infty = P_{Hyd} + P(t) = P_{Hyd} + P_{ac}\sin(2\pi\nu_{ac}t + \varphi) \tag{15}$$

where P_{ac} is the acoustic pressure[2] and φ the phase difference.

Therefore, the Rayleigh–Plesset–Noltingk–Neppiras–Poritsky (RPNNP) equation [Poritsky [40] supplied a modification including the role of dynamic viscosity (η)] can be written as

$$
R\ddot{R} + \frac{3}{2}\dot{R} = \frac{1}{\rho}\left\{
\begin{array}{l}
\left(P_{\text{Hyd}} + \dfrac{2\sigma}{R_{\text{init}}} - P_{\text{vap}}\right)\left(\dfrac{R_{\text{init}}}{R_{\text{fin}}}\right)^{3\gamma} \\[2ex]
+ P_{\text{vap}} - \dfrac{2\sigma}{R} - \dfrac{4\eta\dot{R}}{R} - P_{\text{Hyd}} - P_{\text{ac}}\sin(2\pi\nu_{\text{ac}}t + \varphi)
\end{array}
\right\}
\tag{16}
$$

This equation should be integrated numerically for the bubble behavior to be appreciated (i.e., the temporal response of the bubble radius). However, it must be borne in mind that several assumptions are inherent to obtaining Eq. (16). It is important to approach this aspect because a great deal of effort is devoted to reaching a sensitive description and a better prediction of the dynamics of the bubble, its content, and surroundings. Besides this aspect, it must be noted that Prosperetti and Lezzi [42] demonstrated that the RP equation developed for freely oscillating bubbles remains valid for the description of forced cavities.

2.5 Modifications of the RPNNP Equation: The Role of the Liquid Compressibility

2.5.1 The Approximations

Table 1—which should not be considered as exhaustive—summarizes some of the main assumptions at the root of the RPNNP equation.

2.5.2 Liquid Compressibility

Three kinds of modification have been introduced in order to improve the applicability of the RPNNP equation, and the predictability of the bubble behavior. They concern (1) the compressibility of the liquid, (2) the mechanical behavior of the intracavity thermofluid, and (3) the chemical changes likely to occur inside a collapsing bubble.

From a purely fundamental point of view, the role of the compressibility of the liquid is of major importance. Indeed, an acoustic wave propagates in a fluid at velocity $v = (dP/d\rho)^{1/2}$, where dP and $d\rho$ represent the variation in the pressure and density, respectively. If a medium were incompressible, $d\rho$ would be zero and v, infinity. Table 2 summarizes the modifications associated with point (1).

Table 1. Assumptions Associated with the Obtention of the RPNNP Equation

External medium	Liquid
	Density larger than gas density
	Incompressibility
	Damping due to acoustic radiation is neglected
	Forces acting on a bubble
	No gravity
	Bubble is assumed to be very far from interface (i.e., from other bubbles, solid wall, interface between two immiscible fluids)
Bubble	Radius much smaller than the acoustic wavelength
	Sphericity maintained all the time
	The influence of inward shock or pressure waves is neglected
	Role of the electrical double layer at the bubble interface is neglected
Bubble interior	Vapor content is assumed as constant (no condensation, no evaporation)
	Gas content is assumed as constant
	Intracavity conditions (density, pressure, temperature) spatially constant
	Thermal conductivity is neglected
	Chemical changes are neglected

2.6 Nonlinear Bubble Behavior

2.6.1 Stable Cavitation

2.6.1.1 The dynamics of oscillations. We consider an air bubble (of initial radius R_{init} and bubble wall velocity equal to 0) in water at ambient temperature, for instance. The bubble radius evolves proportionally to this acoustic pressure for very small acoustic driving pressures. This is the linear regime for which there is no sonochemical and sonoluminescent activity.

However, as soon as the acoustic pressure increases, the bubble radius presents a nonlinear response. Such is the behavior of a 2-mm-radius air bubble driven by an acoustic wave with $\nu_{ac} = 10$ kHz and $P_{ac} = 2.7$ bars as predicted by the result of the integration of the RPNNP equation [52] (Figure 6).

At this point it is important to emphasize that 2.7 bars is not a "very small pressure." In fact, the bubble radius is too large for high oscillations to be developed in the conditions quoted above. What is the nature of the problem? A bubble is an oscillator. As a first approximation its response can be assimilated to that of a pendulum [53]. Similarly to a pendulum, the resonant frequency of which is related to the length of its cord, a bubble has a resonant frequency associated with its radius (for an introduction, see [29]). The link between these two parameters (for a liquid with a negligible surface tension) was established by Minneart [54]:

$$\nu_{res} \approx \frac{1}{2\pi R_{res}} \sqrt{\frac{3\kappa p_{Hyd}}{\rho}} \tag{17}$$

Table 2. Common Modifications of the RPNNP Equation Including Liquid Compressibility

Approximations		
Acoustic approximation [43]	The sound speed is considered as a finite constant but the liquid can be assumed as almost incompressible	Valid for predicting bubble behavior in stable cavitation (no dissipation of sound radiation due to fast collapse)
Herring approximation [44]	The sound speed is a finite constant	Valid for transient cavitation provided that maximum radius of a bubble remains a few times greater than the initial radius, only
	The energy stored through compression of the liquid is considered	
Kirkwood–Bethe [45]	The speed of sound is a function of the bubble motion	
Main equations including the compressibility of the liquid		
Trilling [46]	Involves the Mach number (bubble wall velocity/sound speed)	
Gilmore [47]	Involves the differences in the liquid entalpy between the bubble wall and infinity	
	Involves a time-dependent speed of sound in the liquid at the bubble wall	
Keller–Miksis [48]	Incorporates the effect of acoustic radiation by the bubble	
	Involves the approximation of a linear polytropic index[a]	
	The internal pressure is assumed to be constant	
Prosperetti et al. [49]	Involves Keller–Miksis approach modified to take into account more exact formulation of the internal pressure	
	Incorporates an ideal gas law	
Church [50]	Involves Prosperetti and colleagues' analysis but with a van der Waals gas	
Fujikawa–Akamatsu [51]	Involves liquid compressibility; vapor nonequilibrium condensation; heat conduction in the bubble and in the surrounding liquid and temperature discontinuity of the bubble–liquid interface	

Note: [a] It is rare that a process is completely adiabatic (no heat flow) so that the adiabatic ratio ($\gamma = C_p/C_v$) must be exclusively taken into consideration. Moreover it is also rare that a process is fully isothermal (unhindered heat flow). Actually any process is characterized by limited heat flow. Therefore, it is useful to summarize these formulations by the definition of a "polytropic index". The latter (κ) can fluctuate with the heat flow and its value is between 1 and γ (with for instance $\gamma = 1.67$ or 1.40, respectively, for a mono- or a diatomic gas).

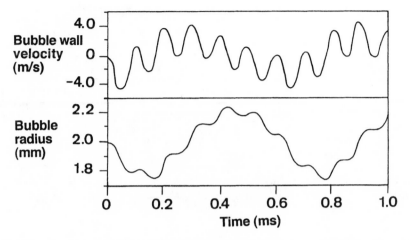

Figure 6. Small nonlinearities in the behavior of a bubble with an initial radius of 2 mm and driven by an acoustic wave with a frequency of 10 kHz and a pressure of 2.7 bars. (After Leighton [52].)

In this equation, ν_{res} and R_{res} are the resonant frequency and radius, respectively, p_{Hyd} the hydrostatic pressure, ρ the liquid density, and κ the polytropic index (see footnote in Table 2). In the case of air bubbles in water at atmospheric pressure, Eq. (17) simplifies to:

$$\nu_{res} R_{res} = 3 \ [\text{ms}^{-1}] \tag{18}$$

In the case of an air bubble in water at ambient hydrostatic pressure, the resonant radius will be 300 μm for an acoustic frequency of 10 kHz. Conversely, the natural frequency of oscillation of a 2-mm-radius bubble will be 1.5 kHz. It can be concluded that in the case quoted above, a bubble with a radius of 2 mm cannot enter the resonant regime (which leads to large radial oscillation) if it is driven at frequencies far from the resonant frequencies. The present reasoning only holds for systems with small departures from linearity. Indeed, as will be seen later, serious divergences from Minneart's law increase as soon as the nonlinear character becomes accentuated in the bubble response. Consequently, it should not be thought that a bubble must reach its resonant (Minneart) radius to collapse violently.

Figure 7 represents the temporal evolution of the radius of an air bubble (initial radius = 100 μm) pulsating in water (conditions: see Figure 6 [52]). This type of response is stable, but its nonlinear character is much more significant than in the case illustrated in Figure 6.

2.6.1.2 Rectified diffusion. Let us consider a bubble of initial radius R_{init} oscillating radially in a sonic field for a long period (i.e., several hundred acoustic

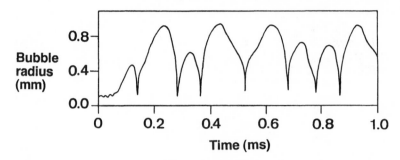

Figure 7. Calculated "radius versus time" curve for a stable bubble (initial radius: 100 μm; conditions: see Figure 6). (After Leighton [52].)

cycles or more) [55]. The vaporization and condensation phenomena likely to occur at the bubble–liquid interface are much more rapid than the bubble dynamics, so that it can be assumed that the vapor pressure inside the bubble will remain constant at equilibrium value.

For the gas the situation is very different. The rectified diffusion mechanism operates in such a way that the bubble must grow. Two contributions, the area and the shell effects, are involved. The area effect is as follows. During compression, the gas inside is at a pressure higher than the equilibrium value and therefore diffuses from the bubble. During expansion, the pressure inside the cavity is lower than the equilibrium value, so that some gas diffuses into the cavity. The flow of gas is proportional to the area of exchange. Since the bubble area is larger during expansion than during compression, there is a net increase in the amount of gas in the bubble over a complete cycle. The shell effect is due to both the proportionality of the diffusion of a gas in a liquid and the concentration gradient of the gas. Let us consider a sphere of constant radius R around a bubble (with $R >$ bubble radius), so that a shell of liquid is defined around the bubble. During bubble expansion the shell thickness decreases. Since the shell is thinner than at equilibrium, the gradient across the shell is higher than at equilibrium. The reverse process occurs during compression. However, expansion is characterized by a larger gradient driving the gas over short distance, while during contraction, a lesser gradient drives the gas over a relatively longer distance. The former contribution dominates.

The conclusion is that a bubble grows in the stable regime [56] as has been demonstrated experimentally.

2.6.2 Transient Cavitation

If the radius of an air bubble pulsating in water decreases more than is described in Section 2.6.1.1, the bubble response changes dramatically as indicated by the predictions derived from the resolution of the RPNNP equation (Figure 8).

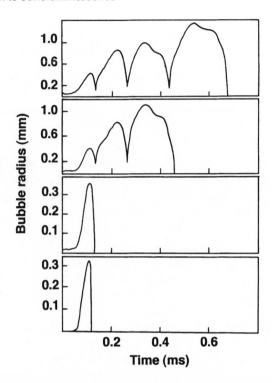

Figure 8. Evolution of the "radius versus time curve" as the initial radius decreases. Acoustic conditions: see Figure 6. (After Leighton [52, p. 310].)

The smaller the radius is, the fewer the oscillations on the part of the bubble. The time scale of collapse is very short and since the expansions are large, the collapses are extremely violent. This is one of the peculiarities of transient cavitation. The reason for such a situation is that the smaller the bubble radius is, the higher its resonant frequency [see Eq. (18)] and shorter the time of response will be. From the point of view of a bubble smaller than resonance size, the expansion time scale seems to be exploded, so that the bubbles have enough time to grow enormously and to store a great deal of elastic energy.

2.6.3 Bubble Behavior: Some Consequences Due to Nonlinearity

The nonlinear character of the link between the radial response of a bubble and the pressure field—as described approximately by the RP equation—is due to term $\dot{R}^2$ (with $\dot{R}$ the bubble wall velocity). A way of illustrating the departure from linearity is to consider a pendulum of length (l) undergoing the effects of gravity (g). Its swing (of angle θ) is governed by

$$\frac{d^2\theta}{dt^2} + \frac{g}{l}\sin\theta = 0 \qquad (19)$$

In the case of a very small swing amplitude (i.e., small θ), $\sin\theta$ can be assimilated to θ. Equation (19) reduces to

$$\frac{d^2\theta}{dt^2} + \frac{g}{l}\theta = 0 \qquad (20)$$

Since the energy dissipation tends to damp the motion, it can only be sustained by an external supply of energy. This can be supplied by means of another oscillator. Under the condition of resonance, both the oscillator and the pendulum oscillate synchronously. However, if length (l) is reduced by factor n (i.e., the frequency increases by factor $\sqrt{n}$), the oscillator will deliver energy $\sqrt{n}$ times less frequently to the pendulum. The consequence is a reduction in the pendulum motion. If the pendulum swings are greater than previously, $\sin\theta$ can no longer be assimilated to θ because $\sin\theta$ increases more slowly than θ. The only way to maintain resonance is to reduce l. Cramer [57] showed that bubbles behave analogously, as indicated in Figure 9.

Figure 9 is still connected with rather low acoustic pressures which are unusual in either SBSL or MBSL. The pattern found by Cramer progressively deteriorates with increasing acoustic pressures as is shown by Prosperetti et al. [49]. For P_{ac} which is likely to induce SL with a single bubble or in stable (multibubble) cavitation, the bubble expansion is greater and greater as the initial bubble radius becomes close to $R_{res}/100$ (Figure 10).

At 20 kHz the resolution of the RP equation modified according to Prosperetti et al. [49] indicates that the most violent type of collapse should be a characteristic of bubbles which start to grow from an initial radius of about 2 µm (by way of comparison, the resonance radius determined via Minneart's law [Eq. (17)] is about 160 µm).

A similar situation occurs at higher acoustic frequencies, as has been determined by calculations by Flynn and Church [59] (Figure 11).

These calculations indicate that in the case of high acoustic pressures like those used for SL (in particular, MBSL), bubbles undergoing large expansion and violent collapse are characterized by initial radii much smaller than the resonance radius. The higher the frequency, the smaller the initial radius leading to large expansion will be. Moreover, at 1 MHz, 80% of the nuclei generated in a multibubble field immediately enter the transient regime [59]. Moreover, in the case of large acoustic pressures and frequencies a bubble may well fail to collapse immediately after the expansion phase. This is due to the fact that the time of collapse is long in relation to the acoustic frequency: The bubble barely enters the compression phase when a new cycle of expansion develops forced by the negative phase of the acoustic wave. This situation is reported in Figure 12.

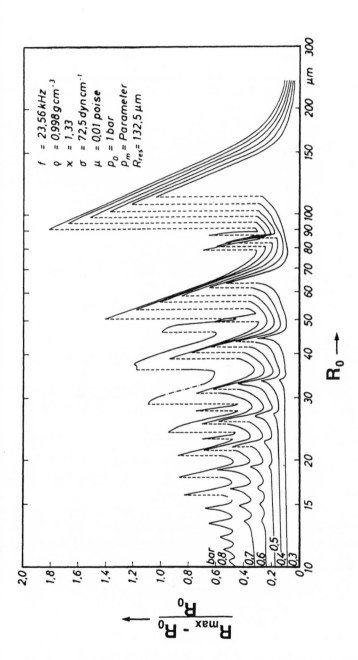

Figure 9. Evolution of the maximum expansion as a function of initial radius R_0 of a bubble. The calculation was carried out for different acoustic pressures (from 0.3 to 0.8 bar). The conditions are reported in the figure itself. Note that the actual resonance holds for $R < R_{res}$ (estimated to be 150 μm according to the Minneart formula [Eq.(17)]. (After Cramer [57].)

THIERRY LEPOINT and FRANÇOISE LEPOINT-MULLIE

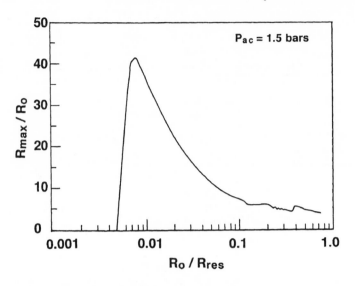

Figure 10. Diagram showing the maximum ratio of expansion (R_{max} is the maximum radius, R_0 the initial radius) as a function of the initial radius. The acoustic frequency and pressure are respectively 21 kHz and 1.5 bars. (After Crum and Gaitan [58].)

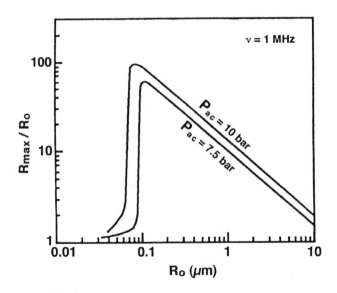

Figure 11. As in Figure 10. P_{ac} is the acoustic pressure and the frequency is 1 MHz. (After Flynn and Church [59].)

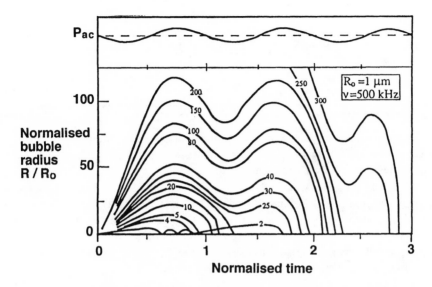

Figure 12. Calculated temporal evolution of the radius of a bubble with an initial radius of 1 μm driven at 500 kHz. The time is normalized as a function of the acoustic period. The number inserted in the curve represents the acoustic pressure/hydrostatic pressure ratio. (After Neppiras [30a].)

2.7 Improving the RPNNP Equation: The Role of the Intracavity Thermofluid

The investigations devoted to the description of the behavior of the intracavity thermofluid belong to two groups, either (1) the role of the thermal conductivity of the intracavity gas (theoretical studies by Hickling [60] and Young [61]) or (2) the mechanical behavior of the intracavity thermofluid and the fluctuations over an acoustic cycle in heat exchanges between the surrounding liquid and the intracavity medium (cf. Nigmatulin and Khabeev [62] and, more recently, Kamath et al. [63]).

2.7.1 An Estimate of the Intracavity Pressure and Temperature

As the bubble compression develops, the temperature, pressure, and density of the intravacity medium increase. In the simple case of an adiabatic collapse the access to the temperature and pressure can easily be calculated as suggested by Noltingk and Neppiras (Section 2.4). The resolution of the RP equation (or its modifications) gives the value of the bubble radius in relation to the time. At each instant, Eqs. (21) and (22) can be applied in order to predict the intracavity temperature (T_{gas}) and pressure (P_{gas}):

$$T_{gas} = T_{init}(\gamma - 1)\left[\frac{R_{init}}{R}\right]^{-3\gamma} \tag{21}$$

$$P_{gas} = \frac{P_{init}R_{init}^{3\gamma}}{R^3 - \alpha^3} \tag{22}$$

In these equations, γ is the polytropic ratio of dissolved gas; R, R_{init}, T_{init}, and P_{init} are the radius at any time, the initial radius and intrabubble temperature and pressure, respectively. Coefficient α is the radius of the van der Waals hard core.

Figure 13 gives the steady-state temporal evolution of the radius and the intracavity pressure and temperature over one acoustic cycle [the case of an air ($\gamma = 1.4$) bubble in water] [64]. The initial radius is 4.5 μm, the acoustic pressure = 1.38 bars, the frequency = 26.5 kHz, the liquid specific mass is 10^3 kg m^{-3}; the surface tension is neglected and the kinematic viscosity is 10^{-6} m^{2}s^{-1}. The minimum radius attained on collapse is calculated as 560 nm (see also Section 3.1.3).

2.7.2 The Effect of the Thermal Conductivity of the Intracavity Medium

In 1963, Hickling [60] refined the calculation of the intracavity temperature by introducing the role of the thermal conductivity of the dissolved gas, i.e., the loss of heat from the bubble into the liquid. Hickling showed that (1) the intracavity temperature reaches a maximum value at maximum collapse, (2) the calculated temperatures are lower than the adiabatic temperatures [for example, a bubble containing Ne at an initial pressure of 0.075 bar (maximum expansion) should be characterized by a maximum temperature of 1500 K instead of 3200 K attained under adiabatic conditions], and (3) the heat loss is larger in collapsing bubbles with smaller radii.

Young [61] derived an analytical expression for the intracavity temperature for a spherical bubble filled with a heat-conducting ideal gas. Figure 14 gives an example of a prediction for the maximum temperature in the central zone of bubbles containing various monoatomic gases. The conditions taken into consideration are as follows. The hydrostatic and acoustic pressures are 1 bar; the initial temperature is 293 K; the radius from which collapse begins is 10 μm and the minimum radius is 3 μm.

These models are, of course, based on many simplifying approximations which make only qualitative predictions reliable. In particular, one of the most significant approximations is that the intracavity fluid is considered to be a gas while, of course, some vapor is present in a real bubble. Even if no trivial answer can be found, the question arises as to the degree of confidence which can be attached to the experimental data, in this way the properties of the dissolved gas are considered exclusively (i.e., its polytropic ratio and thermal conductivity).

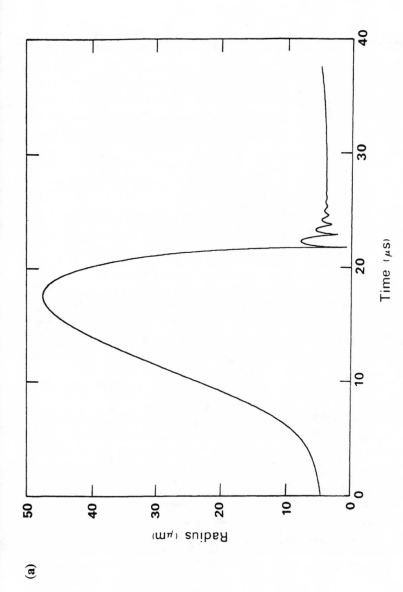

Figure 13. Calculated steady-state temporal evolution of the bubble radius (graph a), intracavity pressure (graph b), and temperature (graph c) over one acoustic cycle of 37 μs duration. (Conditions: see text.)

23

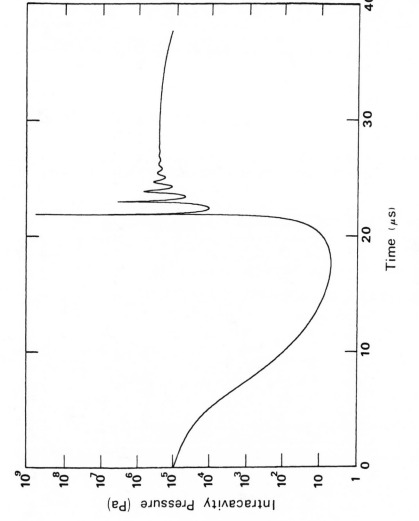

Figure 13. Continued

(b)

24

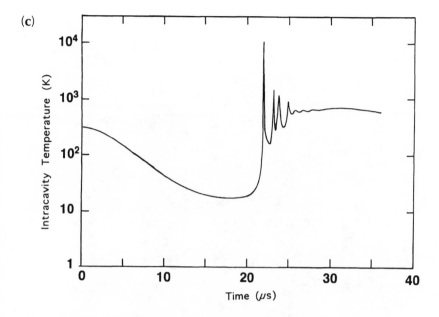

Figure 13. Continued

2.7.3 Nonadiabaticity: Unexpected Consequences

The studies by Nigmatulin and Khabeev [62] and, independently, by Prosperetti's group [49,63] are related both to the thermofluid mechanics behavior of the bubble interior and the liquid adjacent to the bubble–liquid interface. Both studies assume that the gas pressure inside a bubble is uniform (i.e., shock or pressure waves are not considered to form). We report here the essential features of the so-called "complete mathematical model" by Prosperetti's group. If we insist on this approach, it is because it is generally accepted that (1) only the collapse phase is accompanied by heat exchanges from the compressed gas toward the liquid and (2) a thin layer of the liquid surrounding a collapsing bubble heats substantially. The studies mentioned above take into consideration the time scale for expansion and collapse and the large thermal capacity of the liquid.

The basis of this approach is that the duration of expansion is great enough (typically 10 to 20 μs in the case of an acoustic frequency of about 20–30 kHz) for heat transfers to occur from the liquid toward the gas cooled by expansion. The calculation shows that the temperature of the intracavity gas at maximum expansion is close to the liquid bulk temperature. Thus, the initial temperature before the collapse is triggered is higher than predicted by the adiabatic model. Because the collapse is rapid (1 to 2 μs in the conditions mentioned above), the effects of the thermal conductivity of the intracavity medium practically do not operate and the

T (K)

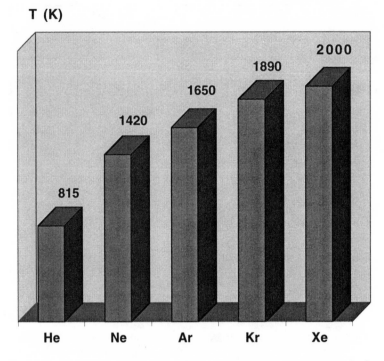

Figure 14. Theoretical estimate of the maximum intracavity temperature in the case of a bubble collapsing from 10 to 3 μm. The temperature at the beginning of collapse is assumed to be 293 K (acoustic pressure 1 bar). (After Young [61].)

bubble interior is made hotter than forecast by the standard "adiabatic" model. Figure 15 shows the steady-state response of an Ar bubble (initial radius 26 μm) in water (acoustic pressure 0.9 bar, frequency 21 kHz). The calculation was carried out over a complete acoustic cycle.

Figure 16 describes the normalized temperature distribution inside the gas at the moment of peak center temperature (upper curve) and maximum radius (lower curve). The corresponding distributions of the temperatures for the liquid outside the bubble are reported in Figure 17.

A detailed description of these "unexpected" effects (i.e., the development of higher intracavity temperature than in the adiabatic case and the modest heating of the liquid just outside the bubble) is given by [63] for the model and by [49] for the set of equations.

Besides this analysis, sonochemists might be interested in the effects of thermal conduction and the latent heat of nonequilibrium evaporation and condensation at the bubble wall under SBSL conditions. Such a recent study—including a large set

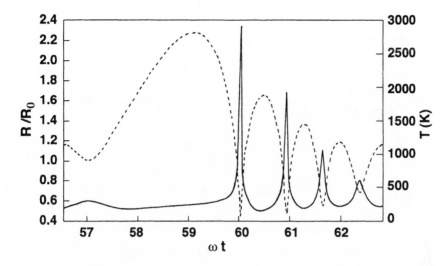

Figure 15. Estimated temporal evolution of the radius (dashed curve) and maximum intracavity temperature (continuous curve) for an Ar bubble of 26 µm initial radius oscillating radially in an acoustic field the acoustic pressure and frequency of which are 0.9 bar and 21 kHz, respectively. (After Kamath et al. [63].)

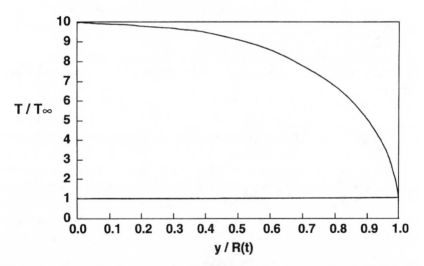

Figure 16. Normalized temperature distribution inside the gas of an Ar bubble (for conditions see Figure 15). The bulk temperature is denoted T_∞; T is the internal temperature; $R(t)$ is the bubble radius at a given time; y is the distance from the center of the bubble. The upper curve corresponds to the peak center temperature and the lower curve corresponds to maximum expansion. (After Kamath et al. [63].) The maximum temperature is of the order of 3000 K and its persistence is of about 100 ns.

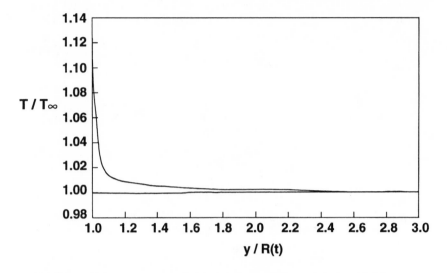

Figure 17. As for Figure 16 but with the liquid adjacent to the bubble–liquid interface. Here, *y* is the distance from the bubble–liquid interface. The upper curve corresponds to peak center temperature while the lower curve indicates the temperature profile at maximum expansion. The predicted temperature increase of the liquid in a shell adjacent to the bubble is ~+40 K. When the dimension of the bubble at maximum compression is taken into consideration (see Figure 15), the temperature gradient is 2.5×10^5 K/mm. (After Kamath et al. [63].)

of chemical equations (the case of an air bubble)—was carried out by Yasui [65]. However, the conclusion of this study has not been totally corroborated by experiment (see Section 4.1).

2.7.4 Supersonic Collapse: The Formation of a Converging Shock Wave

In the preceding paragraphs, it was assumed that the intracavity pressure was uniform on collapse. In the case of an isolated spherical bubble, the resolution of the RP equation (or its improvements) reveals that the speed of the bubble wall may become supersonic (with respect to the speed of the gas inside the bubble) in certain conditions of driven acoustic pressure, initial bubble radius, and frequency. For example, in the case of an air bubble in water with an initial radius of 4.5 μm and driven by an acoustic wave of 1.38 bars and 26.5 kHz (see Figure 13), the calculation shows that the maximum speed of collapse of this bubble may reach about 3000 m s^{-1}.

This led to the hypothesis that provided a bubble remains spherical and its interface smooth on collapse, a perfectly converging shock wave could be launched and could focus its energy to a tiny volume at the very center of the bubble. In this way, the atoms and molecules located in this area could become ionized and a plasma could be generated.

Wu and Roberts [66, 67] and Moss et al. [68–70] coupled the RP equation describing the temporal evolution of the bubble radius with Euler's equation governing the motion of the intracavity gas [71]. Wu and Roberts [66,67] showed that for a number of slightly different conditions of excitation, the response of the system is widely different. If the frequency of the driving acoustic wave is high, the gas in the bubble moves adiabatically ("and no light is emitted" [67]).[3] The calculation shows that as the frequency is reduced (the same for the ambient radius and the driving pressure), the incoming bubble surface acts as a piston that generates an ingoing shock wave that passes through the center of the bubble, and then, when it strikes the bubble surface, halts and reverses its inward motion; a sequence of such an inwardly and outwardly moving shocks occurs. Wu and Roberts [67] say: "The shock waves generate such high temperatures [up to 10^6 K in certain cases of calculation] that the gas near the center of the bubble is almost completely ionized, and emits light, which we attribute to Bremsstrahlung."[4] In this modeling, the duration of the excitation (i.e., the formation of the possible plasma) is extremely short (~ a few picoseconds). The most sophisticated model is due to Moss et al. [68] who used classified "hydrocodes" for calculation (the formation of a converging shock wave is the process by which nuclear physicists currently try to trigger off nuclear fusion via inertial confinement experiments). The main equations and their solutions as well as the mechanism of consecutive shock waves are reported in more detail in [68–70].

2.8 Departures from Sphericity

2.8.1 The Role of the Surrounding Medium

Bubbles in a bubble cloud as well as single bubbles in levitation may undergo various disturbances which are spread over a more or less short time scale and which can lead to severe departures from sphericity.

In the case of a multibubble cloud, any bubble may be disturbed by the presence of a solid wall, so that an inward jet develops until a torus forms (Figures 18 and 19).

Another possibility is that the pressure pulses (Figure 20) emitted by neighboring bubbles in the liquid force asymmetric collapse.

A remarkable experiment directly connected with SL was carried out by Bourne and Field [75] who studied the behavior of an isolated disk-shaped bubble (in gelatin) disturbed by a flat lateral shock wave. The collapse was asymmetric with the formation of a single inward jet. It appears that when the jet hit the opposite bubble wall (Figure 21), a flash of light was emitted (Figure 22a). Then the bubble continued its collapse and the two lobes became luminescent (Figure 22c).

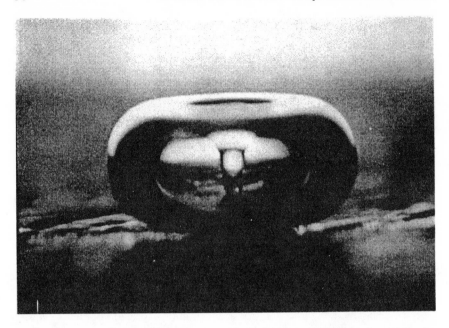

Figure 18. Formation of a jet inside a vapor bubble (in glycerin) forced by the vibration of a plane (frequency: 60 Hz; bubble size: ~2 mm). (After Coleman et al. [72]; reproduced with the permission of L. A. Crum.)

When several bubbles are close to one another, the situation becomes more complex. The interactions between two (laser-generated) bubbles were made by Jungnickel and Vogel [76] (Figure 23).

Theoretical investigations were recently undertaken by Oguz [77], who reported extreme versatility in the behavior of two bubbles close together and oscillating in an acoustic field of 20 kHz. This versatility can be caused by slight changes in the interbubble distance, the radii, and the acoustic pressure. Besides the investigations by d'Agostino and Brennen [78], Chahine [79] has provided theoretical calculations of the behavior of cavities belonging to small bubble clouds undergoing pressure variation arising from nonuniform liquid flows. Figure 24 illustrates the case of a cloud of five identical bubbles. Initially the interbubble distance, the internal pressure, and the size were the same for each bubble. Calculation indicates that the center cavity grows similarly to peripheral bubbles, but terminates with the least distortion.

Of course the situation in an actual field of bubbles is still much more complex and the question arises as to whether spherical bubbles exist in multibubble field experiments. As pointed out by Neppiras [30a], the immediate consequence of such departures from sphericity is a lowering of the maximum internal pressure and temperature via dissipation.

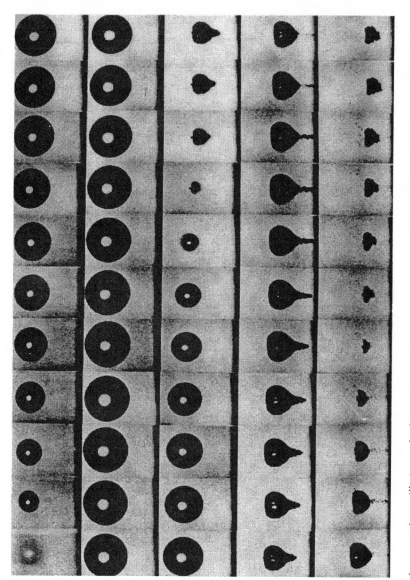

Figure 19. Asymmetric oscillations of a laser-generated vapor bubble in the vicinity of a solid boundary (solvent: water; maximum bubble radius: ~1 mm; number of frames: 75,000 s^{-1}). (Reproduced with permission from Lauterborn and Bolle [73].)

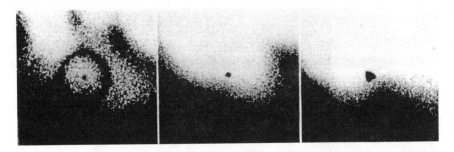

Figure 20. Emission of a shock wave by a collapsed laser-generated vapor bubble (solvent: water). Reconstruction from three frames of a holographic series (30,000 holograms s^{-1}; frame size: 27 mm × 27 mm). (After Lauterborn [74]. Reproduced by permission of Springer-Verlag.)

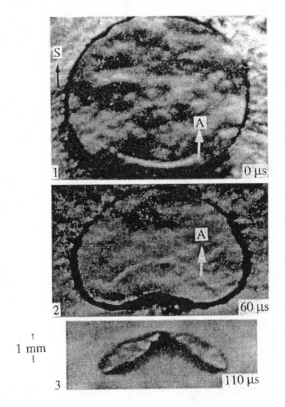

Figure 21. Asymmetric collapse of a cylindrical air bubble (top view) in gelatin. The incident flat shock wave moves as indicated by the arrow S; arrow A indicates the direction of propagation of the "refracted" bowed internal shock wave. The diameter of the bases of the cylindrical bubble is about 12 mm. (After Bourne and Field [75].)

Figure 22. Situation associated with the geometrical distortions reported in Figure 21. (a) Emission of light attributed to the impact of the broad jet on the opposite bubble wall; (b) lack of intermediate emission; (c) secondary emission attributed to the heating caused by the compression of the gas contained in the collapsing lobes. (After Bourne and Field [75].)

2.8.2 The Rayleigh–Taylor Instability

It must be emphasized that departures from sphericity are very common. Even if a bubble may have a spherical shape during expansion and a very large part of its collapse, the rebound phase may be characterized by Rayleigh–Taylor instability. Figure 25 gives the evolution of an ether bubble in glycerin before and after collapse. During the rebound phase (i.e., the phase during which the bubble radius grows again, but generally modestly, see Figure 13), large nonradial oscillations develop which may lead to bubble fragmentation. The origin of the Rayleigh–Taylor instability was introduced by us elsewhere [29]. More advanced descriptions were reported by Plesset and Prosperetti [80].

2.8.3 Coupling the Radial and Translational Motion

In the case quoted above, the bubble was not considered to move translationally. Even an isolated bubble in levitation in a standing wave system undergoes both radial and translational motion. This was predicted by theoretical arguments [82,83] and experimental evidence was recently found [84] (see Section 3.1.4.2). According to both Prosperetti [83] and Longuet-Higgins [82], the immediate consequence of such a coupling may ("and, indeed, must") be the occurrence of an asymmetric collapse.

The origin of the translational motion may be described as follows [29]. We consider a resonant acoustic setup. In the positive phase of the acoustic wave, the pressure is a maximum in the antinodes and zero at the nodes. Therefore, a pressure gradient ($\vec{\nabla}P$ directed—by definition—toward the zones of greater pressure) is set up in the liquid. A bubble of volume V placed in this liquid undergoes a force varying

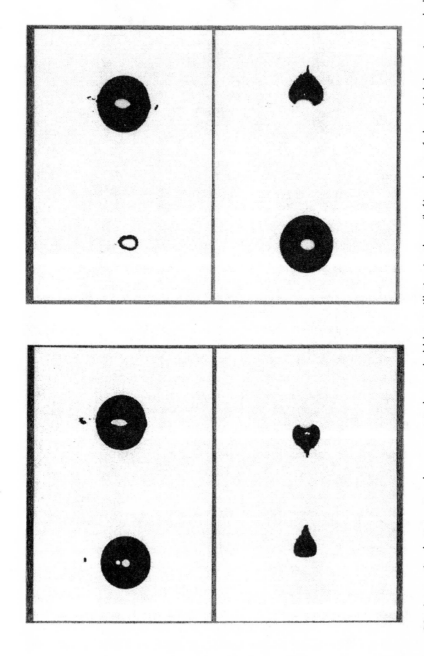

Figure 23. Interaction between two laser-generated vapor bubbles oscillating in phase (left) and out of phase (right); maximum bubble diameter: 2.5 mm. (After Jungnickel and Vogel [76]. Reproduced by permission of Kluwer Academic Press.)

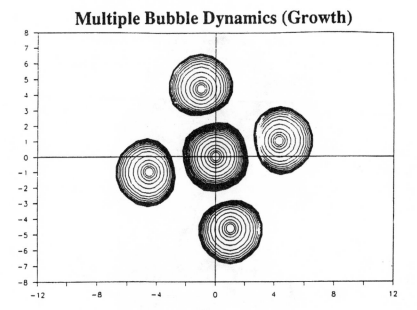

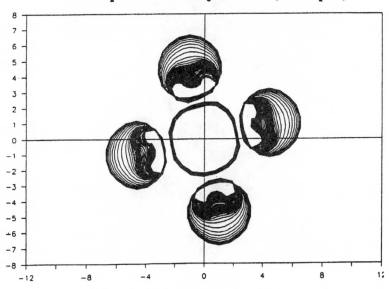

Figure 24. The growth and collapse phases of a cloud of five identical bubbles. The maximum radius/interbubble distance (center to center) ratio is 0.474. (After Chahine [79].)

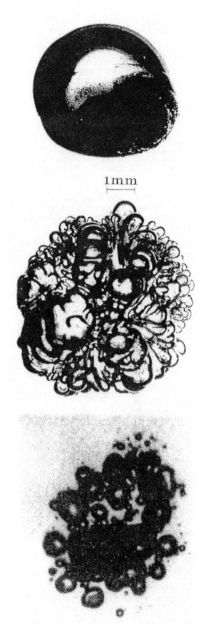

Figure 25. Photographs of an ether bubble in glycerin. Top: before collapse; center: after a collapse and rebound. Bottom: the bubble cloud resulting from a succession of collapse and rebound sequences. (Reproduced with permission from Frost and Sturtevant [81].)

in time since the gradient of pressure is oscillating, i.e., $\vec{F} = -V \times \vec{\nabla}P(t)$. The negative sign expresses the fact that the higher the decrease in the acoustic pressure through the space, the greater the force will be. The latter equation translates the fact that the force is set in the opposite direction to the gradient and that a bubble therefore tends to move in the opposite direction to the local pressure gradient.

Consider a standing wave and a bubble trapped in an antinode (the trapping mechanism is discussed qualitatively by us elsewhere [29]; a quantitative description, based on linear approximation, can be found in Leighton [16]).

Prosperetti [83] says: "Since bubbles tend to move in the direction opposite to the local pressure gradient . . . , in a sound field a bubble drifts toward the pressure minimum (i.e., the antinode) when the pressure falls, and toward the pressure node when it rises. If the bubble is driven below resonance, its volume V expands when the pressure falls, so that the force [acting on the bubble] is greater in magnitude during the expansion than during the contraction phase. Thus, the bubble executes a periodic translational motion in which the upward displacement in the compressed state, under the action of gravity and of the acoustic pressure gradient, is exactly equal to the pressure-gradient-induced downward displacement in the expanded state against gravity."

The bubble radius response to the coupling between the translational and radial motion of a gas bubble behaving isothermally in water was calculated [83]. Figure 26 describes the response of a bubble driven at 1.35 bars ($\nu_{ac} = 26.5$ kHz; the hydrostatic pressure is 1 bar and the surface tension is taken into consideration). Initially the perfectly spherical bubble was at rest with an equilibrium radius of 4.5 µm. In the case of Figure 26, gravity acts downward. The velocity of the inward jet at the impact on the opposite bubble wall could reach about 1000 m s^{-1}.

2.9 Intermediate Conclusions and Comments

Our purpose in Section 2 was to describe the basic tools (drawn from experimental observations) with which physicists and chemists usually try to interpret the experimental data of sonochemistry and SL.

1. It should be remembered that there are three levels of increasing complexity: (1) permanently spherical bubbles with a uniform distribution of the intracavity pressure; (2) spherical (or almost spherical) bubbles with a nonuniformity in their internal pressure caused by the inward propagation of a shock or pressure wave launched via a rapid collapse; (3) bubbles showing nonradial oscillations due to (a) Rayleigh–Taylor instability, (b) the formation of an inward jet because of the presence of a boundary, and (c) the formation of a jet due to a coupling between the radial and transverse oscillation.

In the context of point (1) in particular, more or less strongly simplifying hypotheses are required in order to access the calculations (particularly concerning

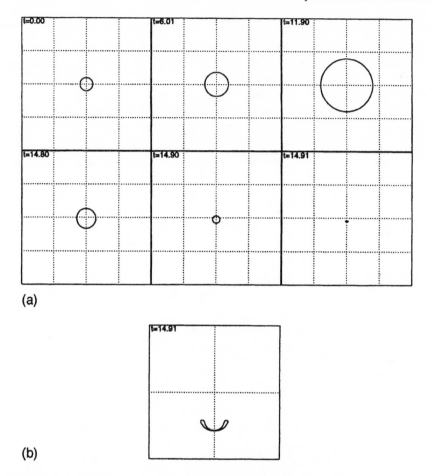

(a)

(b)

Figure 26. (a) Temporal evolution of the shape of a bubble with 4.5 µm initial radius (in unbounded water) placed in a pressure gradient and driven by a pressure field oscillating at 26.5 kHz. Times are in units of 1.1 µs. (b) Blow-up of the last frame of (a). (After Prosperetti [83].)

the internal thermofluid). This limits the field of applicability of some of these equations. In fact, strictly quantitative predictions can be very difficult to make.

2. The short description that we report contains (practically) all of the elements (i.e., the trigger mechanisms) that are invoked by physicists and chemists as the source of SL. Interpretations of the experimental data are based on explanatory models, the most usual of which fall into two broad categories.

A. Several theories describe the activation of the gaseous intracavity phase and its subsequent light emission. This activation can result from the heating caused by the compression (the hot spot theory) or the passage of a (not necessarily spherical) shock wave (the converging shock wave theory).[5] In these cases, the emission phase should involve the intracavity gas and/or vapor (or its debris).

Several descriptions are related to the emission mechanism. For example, some authors assume that the intracavity medium is a luminescent plasma: (1) Wu and Roberts [66,67] and Moss et al. [68–70] consider that light could be emitted by ion–electron bremsstrahlung [88]; (2) Bernstein and Zakin [89] hypothesize that light emission results from electrons trapped in the voids between atoms brought closer together inside the bubble because of the enormous intracavity pressure associated with collapse. In this latter case the degree of ionization would be small ($\sim 10^{-2}$). On the other hand, Frommhold [90] (see also Frommhold and Atchley [91]) suggested that the intracavity medium is practically un-ionized (degree of ionization $< 10^{-4}$) and that the emission results from atom–atom or atom–molecule pairs generated by the passage of a shock wave.

B.   Some models involve the mechanical disturbance of the liquid phase and see the emission coming from the "frozen" liquid in the immediate neighborhood of the interface. In this case the activation may be the cracking of an "ice shell" [92] around a more or less spherical bubble (this shell would form because of the enormous local pressure in the liquid) or the fracture of a supersonic inward jet as it hits the opposite wall of a collapsing bubble. In this latter case [83], the SL emission would be similar to the phenomenon of fractoluminescence [93].

Alternative models were recently proposed because of the stimulation triggered in order to interpret experimental data associated with SBSL (for example, Schwinger [94], Eberlein [95] (see also [96]), Lepoint et al. [64], Garcia, and Levanyuk [97], Prevelenslik [98], Tsiklauri [99], Mohanty and Khare [100], Jauregui et al. [101], Xu et al. [102], Brenner et al. [103], and Vuong and Szeri [104]).

Be that as it may, although it is trivial to mention that SL is affected by the nature of the gas and liquid (as shown via experiments), it implicitly appears from point 2 that there is no consensus on the site of SL activity. This situation is the opposite of what is known in sonochemistry [19]. The fundamental problem of the site of SL emission will be approached further (Section 5.5).

2.10   Investigating SL: Plan and Personal Questions

The first question to be solved is to make sure that isolated bubbles obey the Rayleigh–Plesset dynamics. This question will be examined in Section 3.1.3 (the case of acoustically forced single bubbles maintained far from boundaries). At the present time the problem cannot be solved in a multibubble field.

Since a bubble can distort for a short time near maximum collapse, we will see how bubble shape can be probed in spite of the extreme fugacity of the state of maximum collapse. The analysis concerns single bubbles (Section 3.1.4) and cannot be used for probing bubbles belonging to a bubble cloud.

The third question consists of determining the instant in a bubble's life at which SL flashes are emitted. Though we will discuss the case of a multibubble field, we will focus especially on SBSL (Section 3.2). In this latter case we will look at the way in which the repetitivity and synchronicity of the SL flash were revealed.

The fourth question concerns the energy distribution of photons and the role in the spectral UV-visible distribution (Sections 3.3 and 3.4) of the dissolved gas, the liquid, and some additives.

The fifth question concerns the possible correlation of sonochemical and SL activity. In the case of a multibubble cloud, the question is extremely complex to solve. Taking into consideration the distribution of the bubble radii, it is very possible that SL and sonochemically active bubbles coexist in a given period, so that both activities cannot be dissociated. We will see that single bubbles in levitation afford an unambiguous answer (Section 4).

It will then be possible for some theories on the origin of SBSL to be eliminated (Section 5).

Moreover we will develop an undoubtedly iconoclastic question provoked—among other things—by the preparation of the present text: "is the intracavity gas (or the liquid and/or its debris) the transmitter of SI?" In our opinion, this question has never been solved unambiguously. However, it should not be confused with another question which is more trivial to answer: "is the gas (or the liquid or both) involved in the process of SL?" These items will be looked at in Section 5, i.e., before the general conclusion.

3. SBSL: THE DATA

3.1 Does an Actual Forced Isolated Bubble Obey the Rayleigh–Plesset Dynamics?

3.1.1 *Single Bubbles in Levitation: Experimental Setup and Conditions*

In its most simplified form, an SBSL experiment requires a resonant cell and a correctly degassed solution. The cell may be either spherical or cylindrical and is made of glass or quartz. One or more PZT (lead zirconate titanate: $PbTiZrO_3$) transducers are attached to the cell by means of an epoxy resin adhesive. An oscillating voltage produced by a generator (consisting of a frequency synthesizer and an amplification stage) causes the PZT and therefore the cell to vibrate [22, 106] (Figure 27). A very simple system is described by Putterman [105].

The radius of the spherical cell is approximately 0.03 m for an acoustic frequency equal to 20 kHz. The solution can be degassed by means of an "oil" vacuum pump, so that in the case of water containing air, for instance, the concentration in residual O_2 does not exceed about 3 mg/L. Some gas is introduced into the solution with a syringe or a capillary. A bubble forms "naturally" at the center of the cell because of the primary Bjerknes force (see [29]). A loop of NiCr wire (passing through a

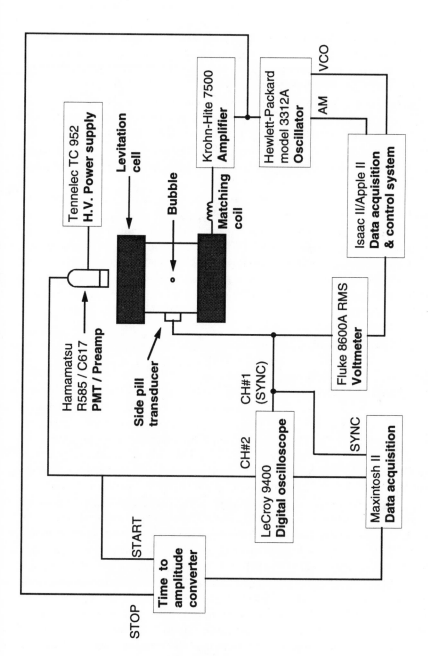

Figure 27. General setup for SBSL generation. (After Gaitan et al. [22].)

cork sealing the neck of the spherical cell) can be immersed at the top of the cell and heated by the brief passage of a low-amplitude electric current. The local heating of the solution causes the vaporization of a small amount of liquid, so that a vapor bubble forms. Through its further radial pulsations this bubble will be attracted toward the center of the cell and turn into a bubble of dissolved gas.

There are several advantages in using a sealed system. First, it becomes possible to prepare solutions with dissolved gases other than air or with volatile additives. Second, a sealed cell can be filled up by solutions, so that there is no free gas/liquid surface likely to cause any disturbance in the stability of the system. Indeed, as will be seen in Section 3.2.1.2, Barber et al. [106], who observed variations in the flash duration and the repetitivity of the flash occurrence, pointed out that "noise increase [could be] caused by the coupling of the sound field to the free surface in the neck of the resonant cell" (i.e., open cell).

In order to avoid spurious reflections of the sound waves off the cell wall, it is necessary to attain a level of resonance. This is done by searching for peaks in the amplitude of the sound field by mean of an immersed needle hydrophone, for instance. Another procedure consists of tracking the phase difference between the voltage and current of the PZT. It must be noted that the resonance frequency is extremely sensitive to the height of liquid in the cell. Moreover, small variations in the ambient temperature require continuous adjustments of the frequency. This can be done automatically with a mode-locking electronic device [28].

3.1.2 Qualitative Bubble Behavior

Typically, a bubble levitation experiment starts with a low acoustic pressure adjusted at ~1.1 to 1.2 atm. This pressure corresponds to an upper threshold for which a bubble is stable against dissolution. The acoustic pressure can be measured via a calibrated hydrophone, the tip of which is colocated with the bubble. Standard guidelines for the correct use of hydrophones (i.e., in order to minimize spatial averaging [16,29]) have been given by Harris [41]. These guidelines take into consideration both the geometrical dimension of the hydrophones and the range of frequencies of interest. For a bubble of an ambient (i.e., initial) radius (R_0) to sonoluminesce in a given liquid with a given concentration of dissolved gas, the acoustic pressure must be increased (physicists commonly say that the parameter space for SL must be reached). Four different regimes can be identified as the acoustic pressure increases [22]. In the dancing regime, the bubble performs small translational amplitude oscillations. For a further increase in the intensity, the bubble develops surface instability and emits small debris which dissolves or is taken up by the main bubble. This is the so-called "shuttlecock" regime, during which a bubble can be maintained for days in spite of its "uncoordinated" movements. In the case of an air bubble in demineralized water, the present authors working with a photomultiplier tube (PMT) able to detect 3×10^{-11} lm did not record any SL signal associated with the shuttlecock regime [107]. At a still higher intensity the bubble becomes stable though without luminescing. The space parameter for SBSL is at still higher drive levels. Under this regime the bubble shrinks in size (the ambient radius decreases; the phenomenon is not explained at the present time [28]). Typical values of R_0 are 4 to 5 μm for an acoustic frequency and pressure equal to 25–30 kHz and 1.3–1.5 atm, respectively. The maximum radius/ambient radius ratio increases up to 10. These values are typical for air bubbles in water. In the case of an ethane bubble in water, Barber et al. [28,108] recorded an expansion ratio equal to 17. In such cases, the collapse duration becomes short (1 to 2 μs) and minimum radii below 1 μm are reached for a few nanoseconds. SL is then clearly visible even in a shady room. In the case of an air bubble in water, SBSL may be maintained for hours or days. The problem of the stability of such a bubble against rectified diffusion (Section 2.6.1.2) is discussed in Section 3.3.2.3. A scheme of the evolution of the bubble behavior as a function of acoustic pressure is given in Figure 28. This qualitative description needs some detailed analyses.

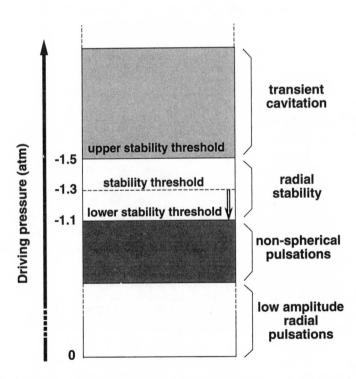

Figure 28. Schematic representation of the different regimes in which a single bubble can be found. (After Gaitan et al. [22].)

3.1.3 Probing the Dynamics of a Single Levitating Bubble

The first study designed to ascertain the temporal evolution of the bubble radius was reported by Gaitan et al. [22] (Figure 29).

The principle of an experiment to measure the radius versus the time (R–t) curve is as follows. A bubble is maintained in levitation and irradiated by the light of a low-power laser. When the bubble contracts (or expands), the intensity of the scattered light decreases (or increases). These variations in light intensity can be recorded by means of a PMT, which produces an output electrical current convertible into a voltage via a resistance. A typical response of such a system is given in Figure 30.

Going from voltage (i.e., scattered light intensity) to radius requires the application of the Mie theory of light scattering [110] which stipulates that in optics, the intensity (I_{scatt}) of the light scattered by a spherical object of radius R is proportional to R^2 for all angles of the scatter. To find the factor of proportionality requires an absolute calibration for at least one known radius. Barber and Putterman [109] achieved this goal by matching a numerical calculation of the hydrodynamic theory of radial bubble motion to that portion of the acoustic cycle where the RP equation is valid (see Sections 2.3 and 2.4). In order to avoid departures caused by the role of compressibility [i.e., near maximum bubble collapse, where the bubble wall velocity may be supersonic (Section 2.5)], they used a modified version of the RP

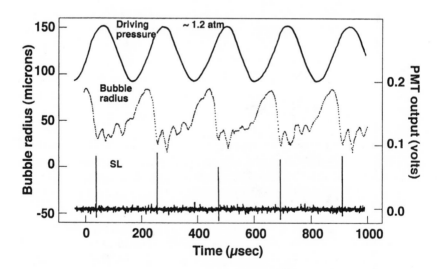

Figure 29. Comparison between (i) the temporal oscillation of the voltage applied to the PZT transducers (upper curve), (ii) the temporal variation of the scattered light by a single sonoluminescing bubble (intermediate dotted curve), and (iii) the trace of the PMT collecting SL photons (lower curve). (After Gaitan et al. [22].)

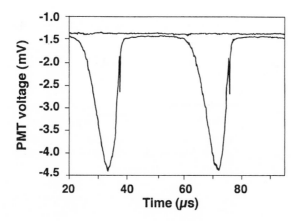

Figure 30. Raw temporal signal obtained by a PMT collecting the photons of the light scattered by a single sonoluminescent bubble irradiated by a low-power laser. Lower voltages correspond to a greater intensity of scattered light, i.e., the bubble in the expansion phase. The narrow spikes correspond to the SL emission. (After Barber and Putterman [109].)

equation from the beginning of the expansion phase to the point where the collapse is approaching (i.e., up to a bubble wall velocity equal to 1/10 the speed of sound in the dissolved gas). The two fundamental adjustable parameters are the initial bubble radius and the acoustic pressure. When an R–t curve [such as obtained in Sections 2.7.1 (Figure 13) and 2.7.3 (Figure 15)] is obtained, the equivalent R^2–t curve can be deduced. The variation of the initial radius and the acoustic pressure is stopped when the calculated R^2–t curve matches—at a factor of proportionality—the experimental I_{scatt}–t curve. The proportionality factor is deduced and the actual R–t curve obtained automatically.

Figure 31 gives the R–t curve for a sonoluminescent bubble [case (a)] and a "bouncing" nonluminescent bubble [curve (b)] [109]. In case (a), the sonoluminescent bubble has an initial radius of 4.5 μm (typically) and its response is calculated for an acoustic pressure of 1.375 atm. The RP equation predicts that if this bubble is perfectly spherical at all times, the minimum radius will be 0.56 μm.

Similar results were obtained by several authors [111–113]. The procedure referred to above enabled Putterman's group to determine the evolution of a bubble's dynamics as the acoustic pressure increases (Figure 32).

In conclusion, in most of its behavior over an acoustic cycle, a single isolated bubble driven by an acoustic wave obeys RP dynamics to a remarkable extent. However, in the region of maximum collapse it might be that the bubble does not approximate a sphere, and if this is so the Mie theory should not be applied. The need arises to discover whether a bubble can distort near the end of implosion since

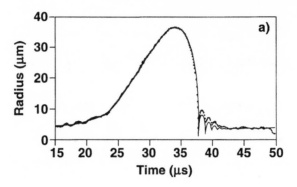

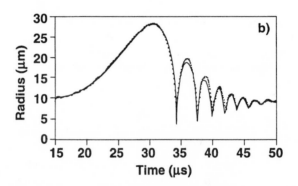

Figure 31. Comparison between the calculated (continuous curve) and experimental (dotted curve) response of (a) a sonoluminescing bubble and (b) a "bouncing" (i.e., nonluminescent) bubble. (After Barber and Putterman [109].)

hydrodynamic theory accepts this possibility (Section 2.8, particularly Sections 2.8.2 and 2.8.3).

3.1.4 Is a Bubble Spherical When It Sonoluminesces?

3.1.4.1 Direct measurements. In order to probe the shape of a bubble at the time of SL emission, Weninger et al. [114] investigated whether the photons were spread out isotropically. Their study involved different cylindrical or spherical resonant cells working between 23 and 40 kHz. Emissions were captured by two PMTs (A and B with the same solid angle of observation), the line of sight of which formed angle θ_{AB}.

If a sonoluminescent bubble was perfectly spherical, PMTs A and B would detect the same intensity. On the contrary, if a bubble was aspherical (ellipsoidal, for instance), the photon density observed in the solid angle of observation would differ

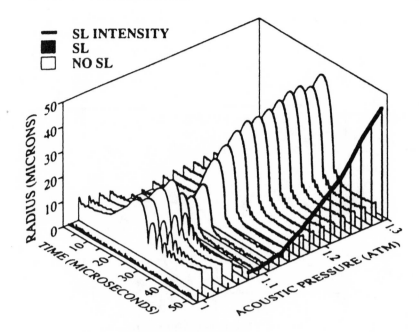

SL INTENSITY
SL
NO SL

Figure 32. Tridimensional graph representing the modifications in the radius versus time curves as the acoustic pressure varies. The case of an air bubble in water (partial pressure in air: 150 mm Hg). (After Barber et al. [28].)

from PMT A or B. Moreover, if the angle between the lines of sight was modified, the photon density observed would also change.

As a matter of fact, it is convenient to use angular correlation (ΔQ_{AB}), which is related to both the total electrical charge recorded by detector A (or B) on the ith flash [$Q_A(i)$ or $Q_B(i)$] and the running average $\overline{Q}_A$ or $\overline{Q}_B$, as given by

$$\Delta Q_{AB} = \frac{1}{\overline{Q}_A \overline{Q}_B} \langle (Q_A(i) - \overline{Q}_A)(Q_B(i) - \overline{Q}_B)\rangle_i \tag{23}$$

where $\langle\ \rangle$ denotes an average over i.

Two fundamental experiments were carried out and two conclusions obtained:

(a) ΔQ_{AB} was measured as a function of angle θ_{AB}. The result is given in Figure 33.

This experiment shows that a bubble is not spherical at the time of emission, and is characterized by a "dipole" component. Two possibilities exist, i.e., the bubble could either be ellipsoidal, or invaded by an inward jet [114]. The latter possibility ties in with the dynamic response described in Section 2.8.3 [82,83]. Of course, the possibility that a bubble may be ellipsoidal at the time of collapse cannot be

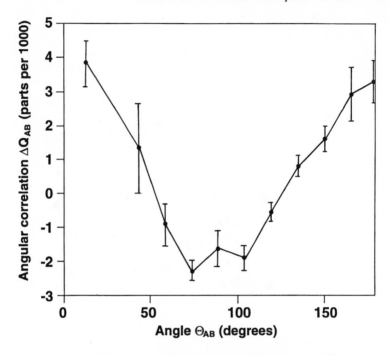

Figure 33. Angular correlation ΔQ_{AB} measured as a function of angle θ_{AB} between two photomultiplier tubes (A and B). If the angular correlation was equal to zero, the distribution in the light emitted by a single sonoluminescent bubble would be isotropic. (After Weninger et al. [114].)

discounted. However, it must be emphasized that ellipsoidal bubbles are known to develop inward jets on collapse as indeed is reported by Prosperetti [115].

(b) ΔQ_{AB} was measured for the red ($\lambda > 500$ nm) and blue ($260 < \lambda < 380$ nm) parts of the radiation from the bubble by using appropriate filters. As is shown in Figure 34, the red correlation is suppressed. This means that longer wavelengths travel across the interface and diffract out of the interior of the bubble on their way to the PMTs. This indicates that the radius (or the equivalent radius, since we are speaking in term of asphericity) of the bubble at the time of emission is about equal to the wavelength of the red light. Note that the RP equation predicts that the mean bubble radius at maximum collapse is 560 nm for a bubble in the condition of SL (Section 3.3). Our point of view is that the diffraction does not necessarily imply that the luminescent region is (exclusively) the bubble interior. Indeed if the mechanism of light were the one suggested by Prosperetti [83] (i.e., fractolumines-cence; Section 2.8.3), the light emitted laterally after the jet impact would travel across the gas phase of the compressed asymmetrical bubble.

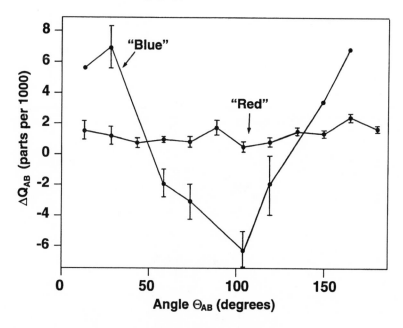

Figure 34. As in Figure 33 but for red and blue photons. Note that there is no angular correlation for red photons. (After Weninger et al. [114].)

This experiment is essentially conclusive as to the fact that a bubble (or more pertinently, the zone of emission) is aspherical at the time of emission.

3.1.4.2 *Probing the liquid flow near an SL bubble.*
In view of Section 3.1.4.1, the need arises to discover the origin of bubble distorsions. In Section 2.8, we relate that various disturbances may lead to departures from sphericity. In order to probe the origin of bubble distortion, Verraes et al. [84] designed an experiment based on a preliminary observation by Lepoint et al. [107], that showed that single bubbles in levitation are chemically active. In this previous experiment a solution of Weissler's reagent [116] was used to yield molecular iodine likely to be trapped with starch so as to form a blue complex. Instead of giving a halo with a more or less isotropic density around a sonoluminescing bubble, this complex developed in the form of a filament (see Figure 57). Using various tracers (sulfur particles produced *in situ* by a levitating bubble, fuchsin spots and dusts naturally present in the solutions studied), Verraes et al. probed the liquid flows in the neighborhood of a luminescent bubble. The flows intrinsic to the presence of such bubbles were clearly identified. A flow with a dipolar[7] character formed around a single SL bubble over a few millimeters (Figure 35). This flow was distorted by the intrinsic flow associated with the propagation of sound in the resonant cells used.

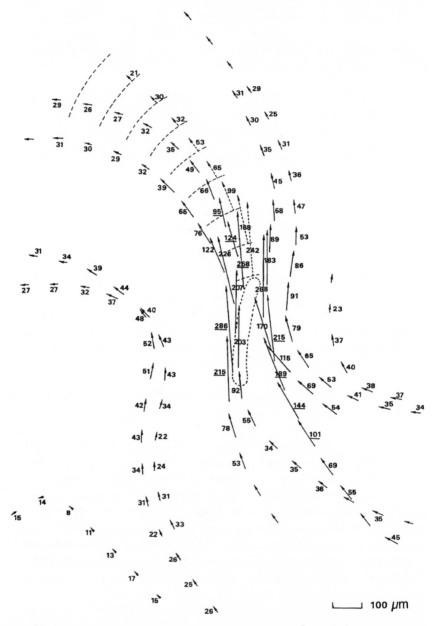

Figure 35. Representation of the distribution of Lagrangian[6] velocities in μm/s of dust (naturally present in the solution) and of sulfur particles generated by decomposition of CS$_2$ contained in water in an experiment involving a single SL and sonochemically active bubble in levitation. The bubble undergoes small frequency (4 Hz) and ample (300 μm) translational oscillation (acoustic frequency 43 kHz) [84].

Such flow patterns reveal that an SL bubble is in a state of radial and translational motion. Indeed, if a bubble were exclusively in translational motion, a particle emitted in the vicinity of the interface would come back to its initial position without drift in relation to the interface. The same is also true for a purely radial motion. As Longuet-Higgins [82] shows from a theoretical point of view, (1) the coupling of radial and translational motion causes drift in any particle initially located near a bubble–liquid interface and (2) this coupling leads to the development around the bubble of a flow with a dipolar character (actually, a "Stokeslet"[7]).

In conclusion, the experiments mentioned above tend to suggest that the origin of bubble distortion—ascertained by the experiments mentioned in Section 3.1.4.1—could be found in the coupling between radial and translational oscillation (see Section 2.8.3). This coupling should result in the formation of an inward jet [82,83]. However, so far it has been difficult to conclude that an inward jet is involved in the process of light emission [83].

3.2 Temporal Characteristics of SL

Numerous studies are devoted to the determination of the temporal characteristics of both MBSL and SBSL. The recent findings on SBSL are extremely interesting because they supply constraints that must involve theoretical explanations (Section 5). Measurements concern the synchronicity of the flashes in relation to the sound period (T_{ac}), the duration of the SL flashes, and the time at which they occur (for example, in relation to the maximum compression).

3.2.1 Single Bubbles

3.2.1.1 Synchronicity and repetitivity. The earlier experiments by Gaitan et al. [22] demonstrated that in the case of single bubbles in levitation, SL flashes occur repetitively near the end of the collapse, i.e., with one flash per acoustic cycle (see Figure 29).

In an advanced study with the fastest available PMT, Barber et al. [108] studied the light emission synchronicity and used a deciliter flask driven by a frequency synthesizer with an intrinsic jitter of 3 ns and working at 30 kHz ($T_{ac} = 33$ μs). They recorded the distribution of the times at which SL flashes were emitted by a single levitating air bubble in water and noted that the best-fit Gaussian curve (occurrence of a flash as a function of the period between two successive flashes) (see Figure 36) had a standard deviation of only 50 ps, so indicating a strikingly high synchronicity of the light emission.

Roy (see [118]) suggested a form of mode locking between a bubble and a resonant acoustic cell as the source of the remarkable periodic stability that Barber and colleagues' experiment reveals. In 1994, Holt et al. [118] used a similar technique to refine these investigations by analyzing the way in which a single bubble emits light when a resonant cell was slightly detuned (resonant frequency: 27 kHz, i.e., $T_{ac} = 37$ μs; detuning over 0.01 kHz). Figure 37 should be considered.

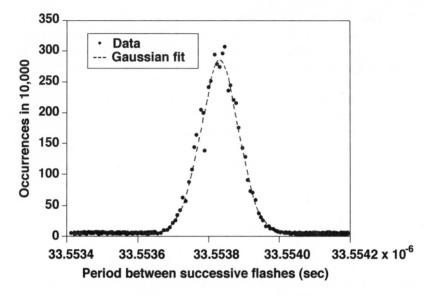

Figure 36. Distribution of the times at which the SL flashes are emitted cycle after cycle. (After Barber et al. [106].)

The experiment consisted of recording a flash at a given time and of delaying the observation via this PMT by 36 µs. After this pause the PMT was switched on (the time is counted from this moment). A flash occurred after a duration Δt which was stored via a computer. The procedure was repeated over 16,000 acoustic periods.

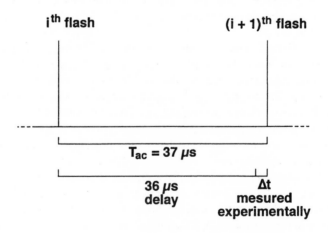

Figure 37. Principle of the method for recording temporal deviations in the occurrence of SL flashes (cycle after cycle).

When the system was correctly tuned, a single-peak histogram (the probability of the appearance of a flash as a function of the time counted after the 36-μs delay in relation to the preceding flash) was recorded, as indicated in Figure 38a. However, when detuning was applied, histograms with several peaks were obtained before a broad distribution was attained (see Figure 38b–d). The successive doubling of the histogram peaks illustrates the chaotic behavior of a single sonoluminescing bubble on frequency detuning. Holt et al. report the analysis of this chaotic behavior in the light emission. However, until now, this approach has not enabled a light emission mechanism to be defined.

3.2.1.2 Flash duration.

One of the most striking measured data on SBSL is the duration of the light flash. Three investigations have been carried out.

In 1992, Barber et al. [106] studied the width of the SBSL flashes under the conditions described in Section 3.2.1.1. The SBSL signal was recorded by means of a two-stage microchannel plate PMT (the rated rise time of which is 170 ps) coupled to a 50-GHz oscilloscope that averaged out the repetitive sampling of the signal. The trace associated with SBSL was compared with the one produced by a 34-ps pulsed laser (Figure 39).

Since the rise curve can be assimilated to a Gaussian law, the authors were able to deduce the flash duration from

$$t_{osb}^2 = t_{PMT}^2 + t_s^2 \tag{24}$$

where t_{osb} is the measured rise time, t_{PMT} the rise time of the PMT (170 ps), and t_s the SL flash duration. With this method, Barber et al. [106] showed that the flash duration could be less than 50 ps. These authors emphasized that "the system is subject to vibration that can, for instance, detune the acoustic resonance. The coupling of the SL to the various sources of noise can therefore lead to variations in pulse height and width that can affect the average trace generated by a sampling scope. Thus methods (e.g., single shot streak cameras) that can record a single flash might possibly find flash widths substantially less than 50 ps."

This remark makes it possible to better interpret the second series of results, i.e., that obtained by Moran et al. [119] who, using a single shot streak camera, deduced that the SBSL flash width could be about 12 ps in the case of an air bubble trapped in a 20% glycerin/water solution at 10 °C.

The third study involved time-correlated single photon counting (TC-SPC) and was recently reported by Gompf et al. [120]. Special attention was paid to the concentration of dissolved gas (the case of an air bubble pulsating at 20 kHz in 250 mL of water). The TC-SPC method, which is suitable for dim intensity, is mainly used in fluorescence to determine decay times. A typical fluorescence experiment may be considered. A laser emitting extremely short pulses (i.e., with an emission time well below the fluorescence decay times) was split into two fractions, one of which excited the sample while the other triggered a time-to-amplitude converter.

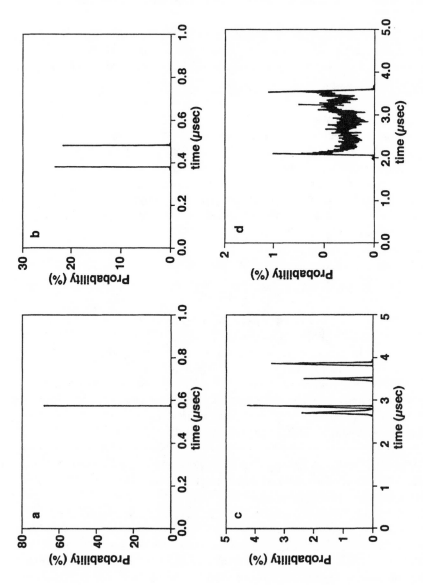

Figure 38. Effect of the detuning in frequency on the time of appearance (Δt) of SL flashes. (After Holt et al. [118].)

54

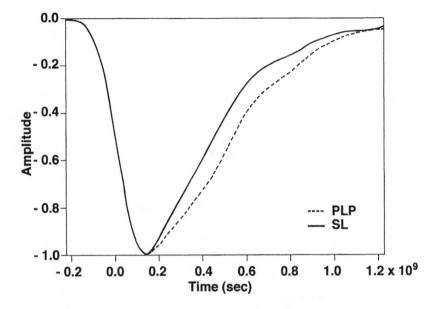

Figure 39. Comparison between the shape of the SL flash at the output of a PMT and the shape of a laser pulse with an intrinsic duration of 34 ps. (After Barber et al. [106].)

A fast PMT was used in order to detect the first photon emitted by the fluorescent sample. The output signal of this PMT served to stop the time-to-amplitude converter. The time so determined was stored in a multichannel analyzer. The sample was reexcited by the pulsed laser and the procedure was carried out over a large number of pulses (typically 10^5). Since the emission of photons by fluoresence is a random process, the time recorded by the time-to-amplitude converter varied from run to run. Over the large number of runs, it was possible to obtain a graph giving the probability of occurrence of times for the observation of a single photon as a function of time. This probability reflected the time-resolved decay of a large population of photons. In the case of SBSL, the first photon emitted by a bubble and detected by the fast PMT served as the trigger of the time-to-amplitude converter and the second detected photon stopped the converter. By way of an example, after the treatment of the signal obtained via the multichannel analyzer, Gompf et al. obtained an almost Gaussian SL pulse with a full width at half-maximum (FWHM) equal to 138 ± 10 ps (the case of an air bubble in water containing 1.8 mg/L in O_2). Repeating the procedure described above for solutions containing different concentrations of air, they showed that the FWHM of the SBSL pulse depended on both the driving pressure and the gas concentration as indicated in Figure 40. However, these key data remain largely unintegrated in an explanatory model, but their refinements (see below) are extremely constraining.

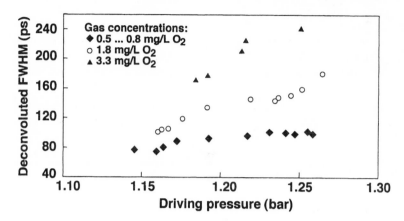

Figure 40. Variation of the SL flash duration as a function of the acoustic pressure and the concentration of dissolved gas (After Gompf et al. [120].)

The method described above was used with a view to determining the width of the SBSL pulses in different zones of the emission spectrum (zone 1: 590–650 nm; zone 2: 300–400 nm). However, no significant differences were noted. As will be seen later (Section 5.1.1), this information, coupled to the very short duration of the flash, enabled the luminescence caused by the compression heating of the intracavity medium to be rejected.

The qualitative observation by Barber et al. that the aging of the water increases the flash duration may be interpreted in the frame of the context of the work by Gompf et al. [120] by referring to the part played by variations in the gas concentration.

Moreover, as suggested by Barber et al. [106, 121], the fact that any increase in the volume of the resonant cell causes the flash duration to increase is probably due to noise increase brought about by the coupling of the sound field to the free surface in the neck of the resonant cell. In this respect the enhancement of the stability of the light emission (mentioned, for example, in [22]) when glycerin was added to water could result from the fact that glycerin damped out any instabilities caused by surface fluctuations in the necks of the resonant cells [106].

3.2.1.3 *Time of appearence of the SBSL flash.*

The determination of the time at which an SL flash is produced by a single bubble was carried out by Barber and Putterman [109] and also by Weninger et al. [122]. In the first series of experiments, Barber and Putterman used the setup described above (Section 3.1.3), which enabled the radius versus time curve to be determined by light scattering measurements. They collected their data by removing the laser line pass filter so that the UV emission due to SL became detectable by the PMT. Taking into consideration the extreme repetitivity of the signal, it was possible to compare the time at which the SL flash occurred with the minimum radius reached by the bubble. Through these experiments it appeared that the SL was emitted about 5–10 ns prior to maximum collapse.

However, in a more recent experiment, Weninger et al. [122] used a more accurate method involving a pulsed laser and determined that the SL flash was emitted within 500 ps of maximum compression.

3.2.2 Multibubble Field

3.2.2.1 Phase of the SL flash.

Stationary sound fields were preferred to progressive waves in order to study the relationship between the SL emission and the sound field phases. This was partially due to the fact that the peak value of a standing wave field is double that of the component running waves [24].

Many preliminary experiments were carried out [123–127] and led to a widespread disagreement as to the SL emission phase. The first reason can be attributed in part to the use of poorly characterized hydrophones and electronics, so that uncontrolled phase shifts were introduced (as indicated by Taylor and Jarman [128]). The second reason was that different experimental conditions were used (the nature and concentration of the dissolved gas, the acoustic pressure and frequency, the hydrostatic pressure), and that before the work by Noltingk and Neppiras [39] (see Section 2.4), it had not been realized that the phase of the acoustic cycle at which a given bubble collapses was not unique (Figure 11). At the present time it is acknowledged that SL emission occurs near or at maximum compression and varies with the parameters mentioned. This aspect is illustrated in Figures 29 and 42. Advanced investigations were performed by Negishi [129], McLeay and Holroyd [130], and Taylor and Jarman [128].

In 1989, Crum and Gaitan [58] clearly demonstrated in another remarkable experiment that bubbles belonging to a multibubble field in the so-called "stable cavitation" regime (21 kHz) emit light with a high level of reproducibility at a fixed phase of the acoustic cycle, as shown in Figure 42a. Knowing the acoustic pressure experienced by bubbles, they determined via the resolution of a modified RP equation the initial radii that bubbles must have in order to collapse at the phase obtained experimentally for SL emissions. As shown in Figure 42b, they observed that the initial radius of SL bubbles must be about 0.01 times the resonant radius (see Section 2.6.3, Figure 10). They showed that (1) the collapse phase angle is essentially independent of the bubble size when only the transient response and the low-radius region are considered and (2) a simple explanation for the uniqueness of the SL emission phase angle is available in that all bubbles able to undergo a transient event give a similar phase angle. Finally they emphasized that strictly speaking, SL from "stable" multibubble cavitation probably does not exist. Indeed,

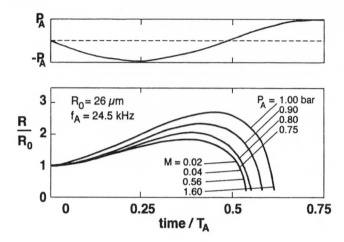

Figure 41. Variation in the time of collapse as the applied acoustic pressure is modified. (After Neppiras [30a].)

bubbles with dynamics as reported in Figure 42b are governed by inertia. This is a characteristic of transient bubbles (Section 2.6.2).

3.2.2.2 Flash duration.   In 1982, Gimenez [7] studied the phase of the sound field at which SL occurs and also noted that the apparent duration of each SL flash (12 ns) was limited by the transit time of the PMT.

More recently, Matula et al. [131] studied the optical pulse width of SL in multibubble fields generated in water containing 80% glycerin, and with air as the dissolved gas. Two systems were used, i.e., an ultrasonic homogenizer, or a PZT transducer attached to the bottom of the cell containing the solutions. The photodetector was a PMT characterized by a rise time of 650 ps. Taking into account the characteristics of the data acquisition system, the resolution was limited to ~1 ns. The SL pulse width measured was limited to 1.1 ns and the rise-time peaks were 650 ps for the homogenizer and 600 ps for the PZT attached to a rectangular cell.

Matula et al. concluded that an upper limit for the pulse width from MBSL of 1.1 ns had been observed and that since this corresponded to the impulse response of the PMT, the actual pulse width of MBSL may be much shorter. In fact, using (1) the experimentally determined response of the system for single photon pulses (~1 ns), (2) the measured width of the signal (~1.1 ns), and (3) the sum-of-squares rule [see Eq. (24)], these authors arrived at the conclusion that the maximum FWHM of the actual pulse width for MBSL is 460 ps. This finding is extremely valuable because it is a strong constraint for MBSL models.

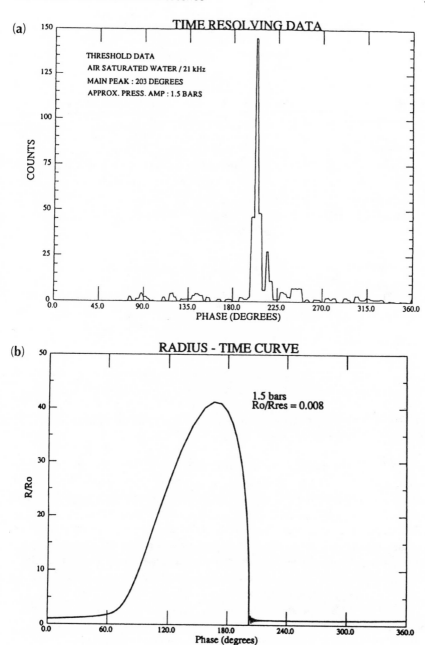

(a) TIME RESOLVING DATA

THRESHOLD DATA
AIR SATURATED WATER / 21 kHz
MAIN PEAK : 203 DEGREES
APPROX. PRESS. AMP : 1.5 BARS

COUNTS

PHASE (DEGREES)

(b) RADIUS - TIME CURVE

1.5 bars
Ro/Rres = 0.008

R/Ro

Phase (degrees)

Figure 42. (a) Phase of SL emission from a "stable" cavitation field (air bubbles in water at 21 kHz and P_{ac} = 1.5 atm). (b) Calculated radius versus time curve for that the maximum collapse coincides with the phase of SL. (After Crum and Gaitan [58].)

3.3 The Intensity of SBSL

3.3.1 Determination of the Number of Emitted Photons

One of the most readily accessible points about SBSL is the number of photons emitted per acoustic flash. This problem was dealt with for the first time by Barber and Putterman [121]. Generally the experiment is as follows. A PMT is set at a distance (r) from the source. The mean value of the electric charge (Q) produced by the PMT can be measured through a 50 Ω resistance, for instance by means of an oscilloscope. The characteristics of the PMT tube are given by the manufacturer, i.e., the gain (g), the first dynode efficiency (E_d), the cathode efficiency (E_c), and the surface of the PMT window (s_w). If q denotes the electron charge, the number of photons (N) is given as

$$N = \frac{Q4\pi r^2}{gqE_dE_c} \tag{25}$$

3.3.2 The Influence of Various Physical and Physicochemical Parameters of Solutions

SL intensity varies as a function of various factors [28] such as acoustic pressure, hydrostatic pressure [132], sudden changes in acoustic pressure, the temperature of solutions, the content and nature of the dissolved gas, the nature of the host liquid, and also the nature and amounts of solutes. These points are reported below. It must be emphasized that we have chosen to adopt a descriptive presentation because the majority of the results do not yet have a reliable interpretation. A discussion of the results will be found in Section 5.

3.3.2.1 The role of the acoustic pressure. Typically, for an air bubble driven in water (20 °C; acoustic frequency and pressure equal to 25 kHz and 1.3 atm, respectively) the number of photons per flash is ~10^5–10^6 in 4π steradians [28].

Working at 10.7 kHz with an air bubble in water, Barber and Putterman [121] recorded the dependence of N on the applied acoustic pressure (Figure 43a) and noted that the SL intensity increased with the acoustic pressure at least until a critical pressure was attained, beyond which the bubble was destroyed. As indicated in Figure 32, a minimum acoustic pressure is also needed to induce SBSL. This minimum pressure depends on the solvent, the content, and the nature of the dissolved gas [28].

Beside these aspects, the number of photons emitted is sensitive to a sudden change in the drive pressure, as Figure 43 (b and c) indicates [133].

3.3.2.2 The role of the temperature of the host liquid. The number of photons emitted per burst by a single bubble is remarkably sensitive to the

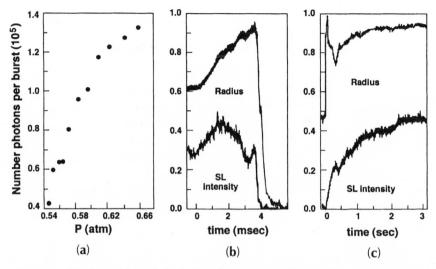

Figure 43. (a) The number of photons per burst as a function of the acoustic pressure amplitude (the case of an air bubble in water; acoustic frequency: 10.736 kHz). (After Barber and Putterman [121].) (b) Dynamic response of the SBSL intensity and the maximum bubble radius to a sudden variation in the drive pressure [the case of an air bubble in water at room temperature (acoustic frequency: 26.4 kHz)]. The acoustic pressure is boosted above the upper threshold and, after becoming brighter for a short time, the bubble disappears. (c) The pressure is boosted from a weak to strong SL under the same conditions. After "gagging," the bubble achieves a new state over a long period measured in seconds. The relative value of the bubble radius is reported for the purpose of comparison. (After Barber et al. [133].)

temperature of the host liquid (Figure 44). Barber and Putterman showed that SBSL occurs over a short range of solution temperatures, typically 0–40 °C for water [133]. Over this range the SL intensity can increase by two orders of magnitude when the temperature is reduced from 40 to 0 °C. This information is crucial. However, it has not yet been (fully) integrated in the explanatory models proposed so far, at least as far as the present authors are aware. It must be noted that in MBSL, numerous studies are devoted to the role of the temperature of the liquid host. Reviews are given in [24] and [27] in connection with the hot spot theory.

3.3.2.3 The role of the dissolved gas. Three aspects can be looked at, i.e., (1) the concentration of dissolved gas, (2) the nature of the dissolved gas and (3) the role of a small amount of monoatomic gas in gas mixtures (the so-called "noble gas doping effect," which appears to apply to nitrogen and/or oxygen bubbles only).

The concentration of the dissolved gas influences the SBSL intensity more or less, according to the nature of the dissolved gas. For instance, in the case of single air bubbles in water (acoustic frequency and pressure: 24.1 kHz and ~1.4 atm)

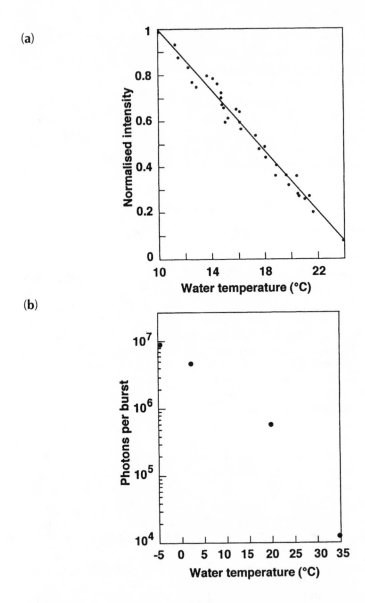

Figure 44. (a) Evolution of the SBSL intensity (visible photons) with the liquid temperature [the case of an air bubble (oxygen concentration: 2.8 mg/L) driven in water; acoustic frequency: 24.1 kHz; acoustic pressure adapted in order to give the maximum possible intensity] (Bertholet et al. [134]). (b) Number of UV-visible photons as a function of the liquid temperature [the case of an air bubble driven in water; acoustic frequency: 27.1 kHz; acoustic pressure: see (a)]. [After Hiller et al. [135].)

Bertholet et al. [134] showed that the SL intensity is hardly affected by variation in the air concentration (Figure 45) at least within a certain range of gas content. This range was determined indirectly, i.e., in the form of the residual O_2 concentration detected by means of an oxymeter ($2.8 < [O_2] < 4.6$ mg/L).

However, the situation is not so simple as Löfstedt et al. [136, 137] and Barber et al. [28,106] report. In fact, there is a rather large range of partial pressures (in the dissolved gas) for which SBSL is known, i.e., ~3 to ~200 Torr. For example, air bubbles are particularly stable over the range of pressures mentioned. Considering such a large range of partial pressures (or gas concentrations), it may appear surprising that bubbles are stable against size increase caused by rectified diffusion (Section 2.6.1.2). In fact, a sonoluminescing bubble should grow during its radial oscillation.

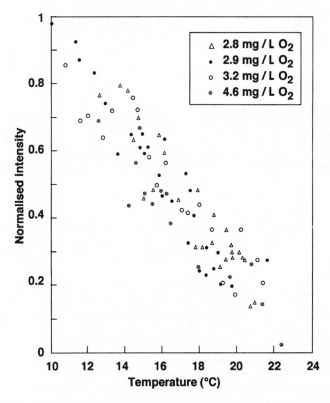

Figure 45. Dependence of the SL intensity (the case of a single air bubble in water) as a function of the air concentration (measured via its oxygen content). Acoustic conditions: see Figure 44a. [After Bertholet et al. [134].)

At low gas concentration, the mechanism explaining the stability against growth involves a steady-state balance of mass flow between the bubble and the gas dissolved in the surrounding fluid. Rectified diffusion cannot operate because in such conditions the diffusion of the gas is the limiting factor. Air and noble gas bubbles obey this diffusion-controlled process in the case of partial pressures of ~3 mm Hg [28,108,136].

At high partial pressure (~200 mm Hg) the behavior of air and noble gas bubbles differs notably [137]. Noble gas bubbles are unstable in that the phase at which SL flashes occur varies, and the SL intensity decreases periodically and then returns to the usual value. Air bubbles are remarkably stable under the same conditions.

The relative instability of noble gas bubbles can be explained as follows. As reported by Barber et al. [28] and Holt and Gaitan [138], a bubble must grow at a high gas concentration. However, at a certain critical size, depending on the acoustic pressure applied, the bubble becomes unstable and loses some gas in the form of tiny microbubbles. The parent bubble is "recycled" and brightly sonoluminesces once again. During the period of microbubble ejection the SL intensity decreases. Since a bubble is no longer in a steady state at high gas concentration, it becomes logical that the SL phase should vary.

Though the study of the region in the parameter space for which SBSL is possible was carried out [138,139], the stability of air bubbles at high concentrations of dissolved air remains elusive. An extra flow of gas should occur during compression. It is conjectured that (1) an outgoing shock wave produced when a bubble is at maximum collapse assists gas ejection or (2) hydrodynamic pulsations of the bubble set up a macroscopic mass convection cell, so enabling the compensation of the diffusional influx [28] to take place.

We can now return to the description of the experimental data [28,108,136,137]. As indicated in Figure 46, the SL intensity depends on both the nature of the dissolved gas and its partial pressure. For instance, hydrogen and deuterium bubbles are stable in water as light emitters only at low partial pressures (~3 Torr). Sonoluminescing ethane bubbles are difficult to stabilize in water: (1) at low partial pressures (3 Torr) they can be maintained in levitation for only 1 min; (2) at partial pressures between 50 and 100 mm the intensity is decreased by a factor of 0.03 in comparison with air bubbles in the same conditions. Such ethane bubbles only survive for 5 to 25 s.

Still greater complexity in the modulation of the SBSL intensity is caused by the addition of small amounts of monoatomic gases. The experiment was carried out by Hiller et al. [140], who used N_2 (99.7% of purity) as the gas dissolved in water. Surprisingly, the SBSL from this type of bubble was much dimmer (~30 times less) than for an air bubble under the same experimental conditions. The same observation was checked in the case of water solutions containing either O_2, or a mixture made up of 80% N_2/20% O_2 (in this latter case, the percentages mentioned do not necessarily reflect the actual composition inside the bubble). The idea occurs that Ar (usually found at ~1% in air) could play an essential role in the triggering of SL.

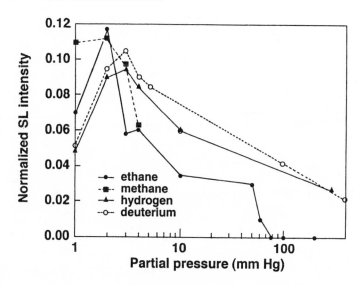

Figure 46. SBSL intensity—normalized to the intensity of an air bubble in water at 20 °C (partial pressure: 150 mm Hg)—for various "pure" gases dissolved at low partial pressure. (After Barber et al. [28,108] and Löfstedt et al. [136,137].)

Indeed, as Figure 47 indicates, small amounts of monoatomic gases (He, Ar, and Xe) dramatically increase the SL intensity of a nitrogen (or oxygen) bubble. However, the effect of noble gas doping does not occur for D_2 or H_2 bubbles.

The spectacular effect of "noble gas doping" is felt to be a key feature in the SL mechanism and is believed to have several causes.

Hiller et al. [140] say: "The key could be the electronic degrees of freedom that account for the radiation. The Penning effect provides an example of an electrical property that is sensitive to minute noble gas impurites. One can also speculate that the origin of the sensitivity could be the result of some process whereby the energy is transferred to electronic degrees of freedom from hydrodynamic motion." However, it must be emphasized that Penning excitation and ionization require long-life (metastable) species that are not very compatible with the short duration of SBSL flashes. Barber et al. [28] also mention that the Penning effect is unlikely.

Recently, Lohse et al. [141] proposed that a bubble could be able to accumulate— specifically—the noble gas atoms. The idea is that a bubble exchanges gas continuously with the solution through the phenomenon of rectified diffusion (Section 2.6.1.2). Lohse et al. consequently suggested that if an air bubble collapses, nitrogen and oxygen could react between themselves or with water vapor. The products so formed would dissolve in the water bulk. Because it is an "inert" gas, argon would not react and would therefore progressively accumulate, so causing an air bubble to turn into an argon bubble (see also Section 3.4.2.2). Until now there is no direct

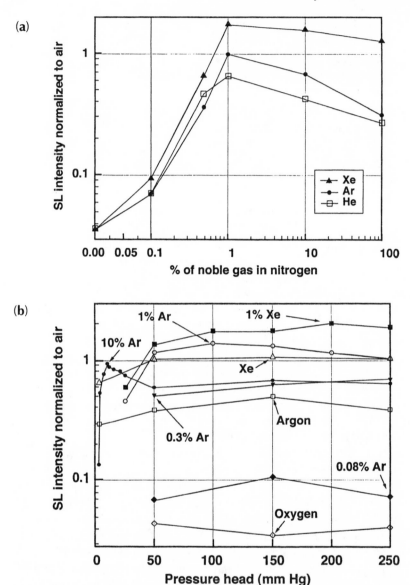

Figure 47. (a) Effect of the addition of several monoatomic gases on the SL intensity of a nitrogen bubble in water. Acoustic pressure: see Figure 44a; acoustic frequency: 33 kHz; temperature of solution: 24 °C; partial pressure in dissolved gas: 150 mm Hg. The unit indicated on the vertical axis corresponds to 2×10^5 photons emitted per burst. (b) Same as (a) but for an oxygen bubble. (c) Same as (a) but for a deuterium bubble (note the absence of noble gas doping effect in this case). [(a) After Hiller et al. [140]; (b, c) after Barber et al. [28].]

(c)

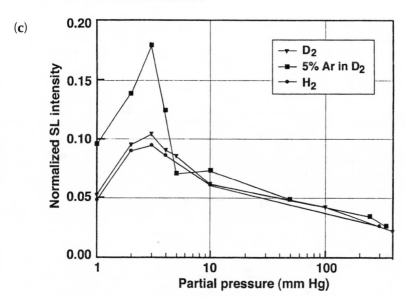

Figure 47. Continued

experimental evidence (though Matula and Crum [142] recently approached the problem) and the origin of noble gas doping on SBSL intensity remains to be confirmed.

3.3.2.4 The nature of the host liquid.
The earlier experiments dealt exclusively with water as the host liquid. Putterman's group showed that heavy water is also suitable for SBSL [143, 144]. However, no phenomenon had been discovered for organic media until recently. In 1995, Weninger et al. [147] showed that low-viscosity silicon oil (Dow Corning 1 cSt), various alcohols, and *n*-dodecane are suitable for SBSL (Figure 48).

One question is to determine whether organic media are as suitable as water for SBSL. In order to establish a strict comparison we must cite an experiment by Barber et al., who mention that in the case of water as the host liquid and of gases at a partial pressure of 150 mm Hg, xenon bubbles emit slightly more than air bubbles ([28, Figure 24, p. 94]). Taking into consideration this remark and the fact that intensities reported in Figure 48 are normalized to 150 mm Hg air in water at room temperature, it can be concluded that in organic solvents SBSL is dimmer than in water.

Of course the high degree of solubility of nonpolar gases in organic liquids has been invoked as a parameter making access to the diffusion-controlled process difficult [28]. However, it must be noted that experiments with alcohols (or

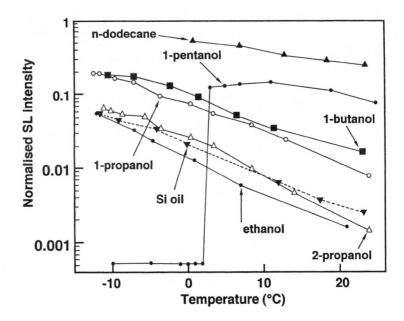

Figure 48. SL intensity from a single Xe bubble trapped in various organic media (the intensity has been normalized to 150 mm Hg air in water at room temperature). The partial pressure in Xe is 150 mm Hg. The duration of SL is >30 s. A relative intensity of 0.05 corresponds to approximately 1.5×10^4 photons per burst. Acoustic frequency: 24–28 kHz. (After Weninger et al. [147].)

n-dodecane and silicon oil) and with water have been carried out with concentration of dissolved gas much below saturation.

As far as the present authors are aware, there is no (clear) explanation to the observation that water is the most friendly liquid for SL as is also pointed out by Barber et al. [28]. In fact, water is the only (known) liquid which accepts extremely stable bubbles, i.e., bubbles with the smallest transverse oscillation, a reproducible phase of emission (for correct tuning), and a stable steady-state intensity for constant experimental conditions.

Is this fact connected with the particular behavior of water known to form clusters with nonpolar solutes (the hydrophobic effect)? In fact, we must acknowledge that the possible connection between the modifications in the structure of a liquid [according to different gas concentration, the nature of the dissolved gas, the role of additive (see below)] and SL has never been explored because it is generally thought that SL occurs mainly in the gas phase. Two notable exceptions deserve to be mentioned: (1) a proposal by Flint and Suslick [145] that (in the case of MBSL in aqueous or alcohol solutions containing alkali salts) the emission from alkali metals could occur in the liquid phase and (2) a recent theoretical work by Prosperetti, who suggests that the SL site is a part of the liquid phase in the immediate vicinity of the bubble–liquid interface [83].

3.3.2.5 *The role of organic additives.*

In 1995, Weninger et al. [147] analyzed the influence of small amounts of organic additives (1-butanol and CS_2) introduced into water. They recorded the intensity of SL from single air bubbles (partial pressure: 150 mm Hg) for various concentrations in these species (Figure 49).

More recently, Bertholet et al. [134] used a sealed cell working at 24.1 kHz to study the effect of three organic solutes (CCl_4, CS_2, and diethyl ether) on the intensity of SL in the case of single air bubbles in water. The aim was to determine whether the volatility of the solutes could directly affect the SBSL process since the results of Weninger et al. [147] tend to suggest that a low-volatility compound (1-butanol; vapor pressure at 20 °C: 4.5 mm Hg) is more disturbing for SBSL intensity than a volatile compound (carbon disulfide; vapor pressure at 20 °C: 280 mm Hg).

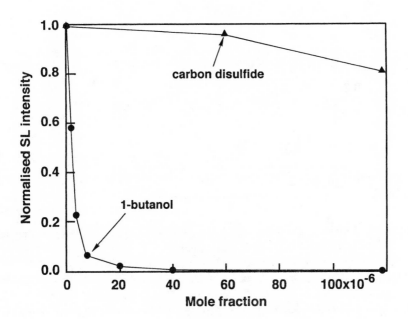

Figure 49. Effect of 1-butanol and CS_2 on the intensity of SL from an air bubble (partial pressure: 150 mm Hg). The intensity has been normalized in comparison with the intensity of an air bubble in water (20 °C; partial pressure of air: 150 mm Hg). Acoustic frequency: 24–28 kHz. (After Weninger et al. [147].)

The results obtained by Bertholet et al. are shown in Figure 50. Since the organic additives were introduced from (not degassed) solutions saturated in the appropriate solutes, Bertholet et al. checked that an equivalent amount of (not degassed) demineralized water did not damage the SL intensity. The vapor pressures (at 20 °C)

(a)

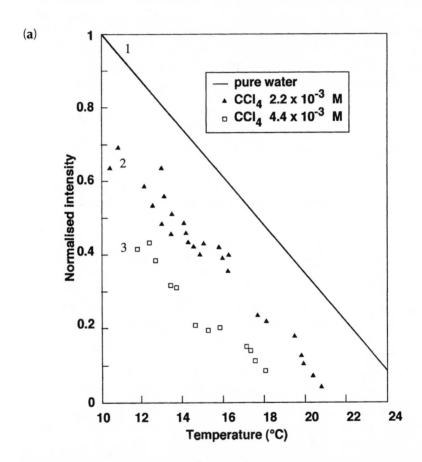

Figure 50. (a) Dependence of SL intensity (the case of a single air bubble) as a function of the temperature of the host liquid (water with various amounts of dissolved CCl_4)(curve 1: demineralized water; curve 2: 2.2×10^{-4} M CCl_4; curve 3: 4.4×10^{-4} M CCl_4) (the oxygen concentration associated with air as the dissolved gas is 3.5 ± 0.5 mg/L). (b) Same as (a) but for CS_2 as the solute (curve 1: demineralized water; curve 2: 2.2×10^{-4} M CS_2; curve 3: 4.4×10^{-4} M CS_2; curve 4: 1.1×10^{-3} M CS_2; curve 5: 2.2×10^{-3} M CS_2) (the oxygen concentration associated with air as the dissolved gas is 3.2 ± 0.2 mg/L). (c) Same as (a) but for diethyl ether as the solute (curve 1: demineralized water; curve 2: 1.6×10^{-5} M; curve 3: 2.2×10^{-5} M; curve 4: 2.4×10^{-4} M) (the oxygen concentration associated with air as the dissolved gas is 3.2 ± 0.2 mg/L). (Bertholet et al. [134].)

(b)

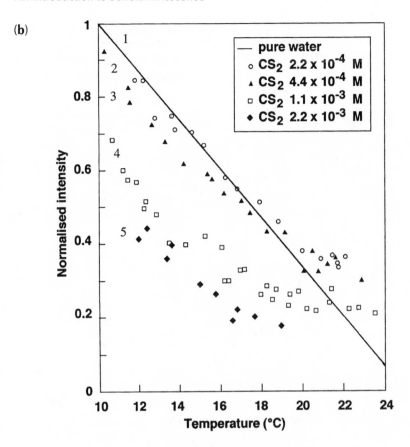

Figure 50. Continued

were 85 (CCl$_4$), 280 (CS$_2$), and 420 mm Hg (diethyl ether). There is therefore no strict correlation between SBSL intensity and the vapor pressure of solutes. Moreover, the results obtained by Bertholet et al. [134] on the role of CS$_2$ and diethyl ether tend to suggest that for low concentrations these additives are able to improve SL activity, at least within a certain temperature range in the host liquid.

3.3.2.6 *The case of single Xe bubbles in various alcohols: sudden transitions due to temperature.* Weninger et al. [144] report a curious phenomenon in their study related to SBSL in organic media. This phenomenon—which concerns Xe bubbles—can be observed for 1-pentanol (Figure 48) and 1-butanol (Figure 51). The SBSL intensity suddenly increases at a "critical" temperature which depends on both the solvent and the concentration and nature of the solute. This situation is rather reminiscent of phase transition processes known to occur

(c)

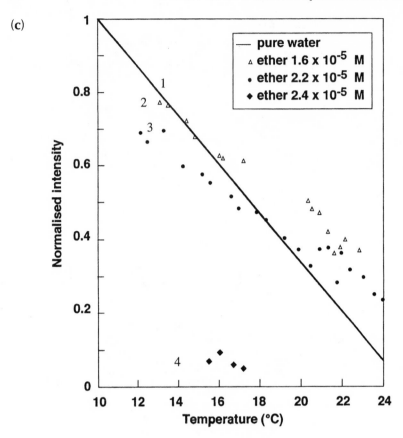

Figure 50. Continued

when a "solute–solvent" structure is modified, for instance. However, the data are fragmentary and at the present time, their significance still remains elusive.

3.3.2.7 A summary of the data obtained up to 1997 on the variation of SBSL intensity See Table 3.

3.4 Spectral Data

3.4.1 Experimental Setup: A Comment

As soon as the SL intensity is affected by physical and physicochemical factors (Section 3.3), it becomes necessary to determine whether this change is accompanied by significant modifications in the spectral distribution. This question was

Table 3. Literature Related to the Variation of SBSL Intensity

Liquid	Gas	Gas P (Torr)	Freq. (kHz)	T_{sol} (°C)	Remarks	Ref.
Water + glycerin	Air	<760	10 to 30		P_{ac} chosen to maximize light output	106
Water	Air	<760	27	3 to 22	At each T the brightest light output recorded	133, 135
Water	x% He in N_2	150	25	24	x = 0, 0.1, 0.5, 1, 10, 100% intensities are normalized to air (~2 × 10^5 photons)	28, 146
	x% Ar in N_2	150	25	24		28, 146
	x% Xe in N_2	150	25	24		28, 146
Water	N_2, Ar	50 to 250	24	24	Intensities are normalized to air	28, 146
	x% Ar in N_2	1 to 480	24	24	x = 0.1, 1, 10%	28, 146
Water	O_2, Ar, Xe	150		25	All SL intensities normalized to air	28
	x% Ar in O_2	150		25	x = 0.8, 0.3, 1, 10%	28
	x% Xe in O_2	150		25	x = 1%	28
Water	C_2H_6	1 to 80	35		Intensities are normalized to air at 150 Torr	108
	D_2	1 to 400	35			108
	D_2/5% Ar	1 to 400	35			108
	N_2/1% Xe	50 to 500	35			108

(continued)

Table 3. Continued

Liquid	Gas	Gas P (Torr)	Freq. (kHz)	T_{sol} (°C)	Remarks	Ref.
n-Dodecane	Xe	150	24	−10 to 25	Intensities normalized to 150 Torr in water at room T	147
Ethanol	Xe	150	24	−10 to 25	Idem	147
1-Propanol	Xe	150	24	−10 to 25	Idem	147
2-Propanol	Xe	150	24	−10 to 25	Idem	147
1-Pentanol	Xe	150	24	−10 to 25	Idem	147
1-Butanol	Xe	150	24	−10 to 25	Idem	147
Si oil	Xe	150	24	−10 to 25	Idem	147
1-Butanol	Xe	3 to 450	24	−10 to 25	Idem	147
H_2O + 1-Butanol	Air	150	24	25		147
H_2O + CS_2	Air	150	24	25		147
Water	Air		26		Dynamic responses to I_{SL} to sudden jump in I_{sound} $t = 0$ to 3 s	133
Water	Ar, N_2	110	30	10, 25	Light emission evolution with rapid increase in drive P and after being in steady state	142

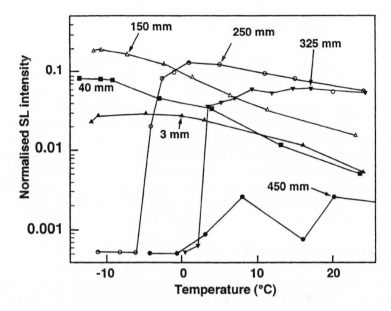

Figure 51. The intensity (normalized to a 150 mm Hg air bubble in water at room temperature) of SL for a single Xe bubble in 1-butanol as a function of the temperature of the host liquid and the partial pressure. (After Weninger et al. [147].)

solved via the recording of UV–visible emission spectra. The wavelength range is limited by the absorption of the host liquid (and possibly its additives) and the spectral window of the spectrograph (e.g., the type of detector, grating, optical fiber).

It is important to emphasize that the use of optical fibers coupled to a focusing system specifically adapted to a monochromator ensures the best result in spite of an attenuation of the signal. Indeed, such systems enable the image of the source to be reconstructed at the entrance slit of the monochromator, so that both internal mirrors and the grating are correctly illuminated. If lines and bands are present in a spectrum, the only instrumental disturbance afforded these spectral features is the instrumental broadening caused by diffraction at the slit and at aperture stops, lens errors, and imperfections in the mirrors and the other optical parts inside the spectrograph. This broadening may be determined and quantified by means of the line of a laser or a low-pressure lamp.

While MBSL spectra are characterized by lines, bands, and broad continua [27], SBSL spectra only exhibit broad continua extending from ~700 to (probably) less than 200 nm whatever the nature of the liquid and the gas (resolution of the spectrograph: 3 nm) [28]. This is also true for aqueous solutions of NaCl, while in MBSL, it has been established that the lines of the alkali metals are detectable when solutions of the corresponding salts acoustically cavitate [148] (see also Part B of Table 4).

3.4.2 Water as the Host Liquid

3.4.2.1 The role of the host liquid temperature on the spectral distribution. We will only discuss the SBSL spectra. Parts C and D of Table 4 (Section 3.4.4) itemize the data obtained up to 1998. However, in order to help the reader go further in MBSL, we summarize the MBSL data in parts A and B of Table 4 (Section 3.4.4). In this latter case, we consider only the period from 1957, for which significant progress in the field of spectral resolution was made.

Figure 52 [28] shows the spectral distributions associated with single He bubbles (partial pressure: 150 mm Hg) in water at various temperatures (0–25 °C). It can be concluded that the spectral distributions are the same (only the light intensity decreases with the increasing temperature of the solution). It can therefore be deduced that the number of light transmitters diminishes, but that their nature is not modified.

It must be emphasized that solutions containing monoatomic gases other than He behave similarly in relation to the temperature effect [28]. It should be noted that the response of MBSL with respect to the temperature of the host liquid is similar to what is observed in the case of SBSL, i.e., a decrease in intensity without any measurable variation in the spectral distribution (see the work by Seghal et al. [149] and by Didenko et al. [150] in Figure 53).

3.4.2.2 The role of the dissolved gas. The role of the dissolved gas in the spectral distribution of SL from single bubbles in water was studied by Putterman's

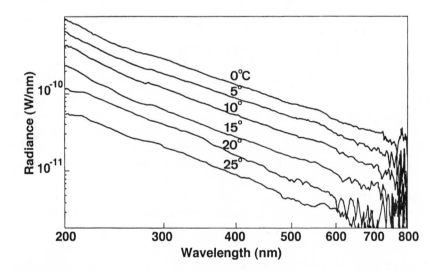

Figure 52. Corrected spectra for a 150 mm Hg He bubble in water at various temperatures. (After Barber et al. [28].)

group [28,135,140,147]. An example of this effect is given in Figure 54a in the case of water solutions with different dissolved noble gases.

Two points can be deduced from Figure 54a.

(1) In spite of the absence of correction for the transmission, it appears that the spectral distribution changes for the different monoatomic gases (this fact is also well known in MBSL), particularly for $\lambda < 320$ nm. This means that the light transmitters are not the same (or behave differently) as soon as the dissolved gas changes. However, it is difficult to certify that the gas is the SL transmitter. Indeed, as Barber et al. [28] mention especially, the spectra do not reveal the presence of lines or bands associated with the gas (this remark also applies to MBSL).

(2) It can be deduced that the intensity of SBSL (the case of monoatomic gases) increases with the size of the atoms of the monoatomic gas. The gas is clearly involved in the global process of SBSL (this is also known in MBSL).

However, a large number of physical and physicochemical parameters are associated with the size of monoatomic gases [e.g., ionization and first excitation potentials, thermal conductivity, polarizability, the free Gibbs energy for clusterization in connection with the "hydrophobic effect" (the case of water as the solvent for nonpolar gases)]. The experimental observations reported above do not enable

(a)

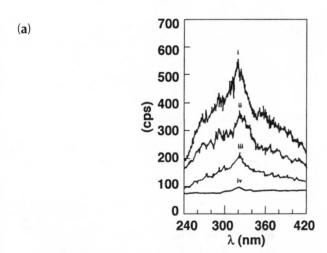

Figure 53. (a) Rapid scans of the low-resolution spectra of neon-saturated water (240–420 nm) at various bulk solution temperatures (i) 25 °C; (ii) 32.5 °C; (iii) 52 °C; and (iv) 69 °C (MBSL at an acoustic frequency of 459 kHz). (After Seghal et al. [149].) (b) MBSL spectra of Ar-saturated water (acoustic frequency: 22 kHz): (i) 11 °C (intensity divided by 2); (ii) 12.5 °C; (iii) 26 °C, and (iv) 48 °C. (After Didenko et al. [150].) (c) MBSL spectra of Ar-saturated water (acoustic frequency: 1.1 MHz): (i) 11 °C; (ii) 30 °C; (iii) 39 °C, and (iv) 50 °C (After Didenko et al. [150].)

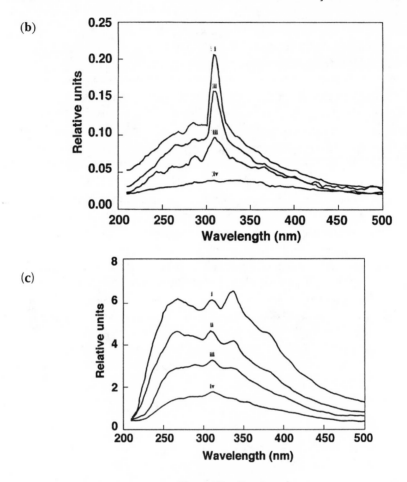

Figure 53. Continued

the parameter actually involved in SL to be specified and, from this point of view, are not constraining enough for a theory to be postulated.

From Figure 54b, a remark emerges which concerns the role of the "noble gas doping" effect. It should be noted that the spectral distributions associated with SL from an aqueous solution of 1% Ar in nitrogen and an aqueous solution of Ar (Figure 54b) are very close. However, it must also be acknowledged that the situation is very different when He is involved in analogous experiments, as is indicated in Figure 54b. Consequently, supplementary data should be obtained in order to provide experimental evidence in favor of the chemical rectified diffusion mechanism mentioned in Section 3.3.2.3 [141] (see also Matula and Crum [142]).

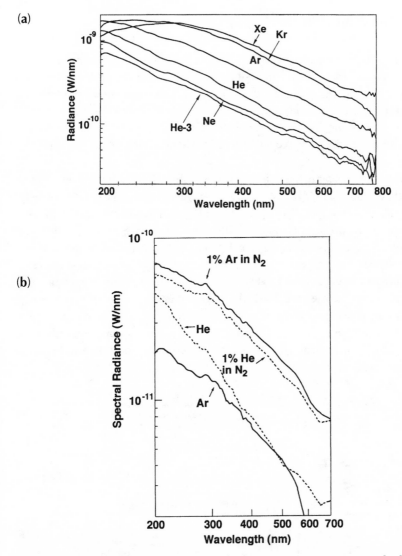

Figure 54. (a) Spectra of SBSL in water for various monoatomic gases [28]. These spectra are not corrected for transmission in the liquid. The partial pressure in gas is 3 mm Hg. (b) Comparison between the spectra associated with bubbles floating in water at 24 °C; dissolved gas (at partial pressure equal to 150 mm Hg): (i) 1% Ar in nitrogen, (ii) Ar, (iii) 1% He in nitrogen, and (iv) He. The data in (b) are corrected for the absorption of water and quartz, and for the voltage dependence of the quantum efficiency of the photodetector. In (a) and (b) the acoustic pressure is not mentioned; the acoustic frequency is 33 kHz. The resolution of the spectra is 10 nm. A recent study by Hiller and Putterman [151] was carried out at a resolution of 3 nm but did not reveal any measurable changes (i.e., lines or bands).

3.4.3 Organic Media as the Host Liquids

As a transition from the preceding paragraph it must be mentioned that the emission spectrum associated with a single bubble of ethane in water at 20 °C and with a 150 mm Hg Xe bubble in n-dodecane at –2 °C were recorded by Weninger et al. [147] (Figure 55). In both cases the Swan bands [152] due to the emission of excited C_2 are not detectable even though the resolution is 10 and 3 nm for the spectra in Figure 55. Single Xe bubbles in 1-butanol or ethanol behave similarly [28,147].

It is important to bear in mind that Flint and Suslick [153] showed that C_2 bands are clearly observable in MBSL (ν_{ac}: 20 kHz, solvent: n-dodecane or low-viscosity silicone oil, dissolved gas: Ar) as indicated in Figure 56a. The fact that SBSL and MBSL spectra are different could lead to the conclusion that SBSL and MBSL are two different processes. From a general point of view it is known that a given process (a shock wave, for example) may lead to the emission of lines superimposed on a continuum in the case of weakly dense media, and that only a continuum is observable for very dense ones [154]. Moreover, Weninger et al. [155] very recently recorded the emission spectrum of a single 290 mm Hg Xe bubble near a solid boundary (solvent: n-dodecane). Both a continuum and the Swan bands (emission of excited dicarbon molecules) are observable (Figure 56b). The difference between MBSL and SBSL seems to be less than initially thought.

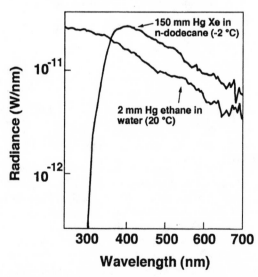

Figure 55. Spectral distribution of a single ethane bubble in water at 20 °C and of a 150 mm Hg Xe bubble in n-dodecane at –2 °C (After Weninger et al. [147].)

(a)

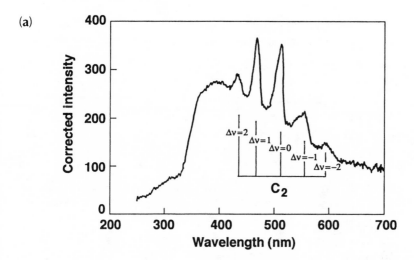

(b)

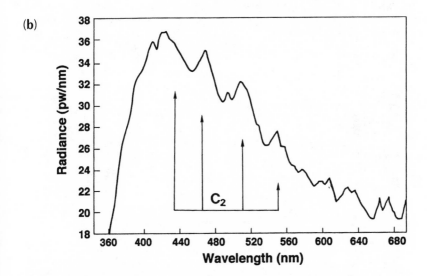

Figure 56. (a) Spectral distribution of a multibubble cloud (Ar as the dissolved gas; *n*-dodecane as the solvent). (After Flint and Suslick [153].) (b) Spectral distribution of a 290 mm Hg Xe bubble in *n*-dodecane. The bubble is near a solid boundary. (After Weninger et al. [155].)

3.4.4 A Summary of the Investigations in Which SBSL and MBSL Spectra Were Recorded (*See Table 4*).

Table 4A. MBSL Spectra in Water

Liquid	Gas	Freq. (kHz)	Cont.	Cont. Max. (nm)	Peaks: Bands or Lines (nm)[a]	Rem.	Ref.
Water	He	337	s (strong)	270–340	246, 262, 293, OH* (**310**, 340), 365, 407		156
	Ne	459	s	280–340	OH* (280, **310**, ~340), ~380, ~445	≠T°	149, 157
	Ar	16	w (weak)		OH* (**310**)		128
		22	s or m	270–330	~270, OH* (~285, **310**)	≠T°	150, 158, ≠I$_{US}$
		300		125, 350	No peaks		159
		333	m (medium)	300–350	OH* (281, **310**, ~340), cont (H$_2$O*,OH*,H+OH)		157
		337	s	260–290	268, OH* (280, **310**, 340), ~380	≠T°	150,156, 158, ≠I$_{US}$
		459	s	300–350	~270, ~290, OH* (**310**), ~350		157
		500	s	300–350	OH* (**310**)		158, 160
		800	s	≤450	H$_2$O vap?: 615, 582, 550, 512, 498, 476, 452, 430		128
		863	s	~250	**270**, OH*(280, **310**, 340), cont(H$_2$O*,OH*,H+OH)		161
		1090	s	260–350	~270, OH* (~280, 310, **340**), ~385	≠T°	150
	Xe	16	m	300–350	OH* (**310**)		128
		300		126, 350			159
		337	s	~240	**240**, OH* (280, 310, 340)		156
		500	s	300–350	OH* (310, 340)		128
	Kr	337	s	~250	250, ~270, OH* (280, 310, 340), ~380		156
		459	s	290–370	OH* (~310, ~**340**)		157
	Air	16	s	~370	OH* (**310**), 370		128
		24	s		~900, ~1000	IR em.	162

Table 4A. Continued

Liquid	Gas	Freq. (kHz)	Cont.	Cont. Max. (nm)	Peaks: Bands or Lines (nm)[a]	Rem.	Ref.
		459	s	300–440	OH* (~310), ~360, ~430		157
		500	s	310–350	—		128
		1100	s	260–350	270, 318, 340		150
	N_2	459	s	330–350	OH* (~310), ~350		157
	O_2	459	s	320–340	max ~310		157
		800	s	≤450	H_2O vap?:615, 582, 550, 512, 498, 476, 452, 430		161
	NO	459	m, s	280, 400			163
	NO_2	459	s	~400		(Resol. Å)	163
Water + NaCl or NaI	Ar	20	m	≥300	OH* (310), Na* (**589** unresolv.)	MBSL	164, 165
		32	s	≥300	—	SBSL	165
		460			554, Na* (589.6, 590) no shift	8 & 40	166
	Xe	16, 20	w		Na* (589 unresolv., shift ~10 Å, hwm ~30 Å)	low	128
		294			Na* (589.6, 590, no shift) + tail (590–596)	0.6	167
		500	m		Na* (589 unresolv., shift 10 Å)	low	128
	Kr	500	s	~315	~315, ~370, ~560, 589	low	128
H_2O+KI or KOH	Ar	20			K* (767, 770) redshift	2	145
		460			740, K* (767.8, 771.1) redshift 13 Å, ~795	8 & 40	166
H_2O + HNO_3	Ar	459	s	~350	Absorption by HNO_3 ~265–335 nm ≠quenching		168

(continued)

83

Table 4A. Continued

Liquid	Gas	Freq. (kHz)	Cont.	Cont. Max. (nm)	Peaks: Bands or Lines (nm)[a]	Rem.	Ref.
H_2O + XNO_3	Ar	459	m	~350	X = Li, Na, K		168
H_2O + $CuSO_4$	Ar	459	s, m	333, 527			169
H_2O + I_2	Ar	459	m, s	270, 332		40	170
H_2O + Br_2	Ar	459	w, s	249, 367		40	170

Note: [a]Most intense in bold; not given by author but deduced from spectra in italic.

Table 4B. MBSL Spectra In Nonaqueous Liquids

Liquid	Gas	Freq. (kHz)	Peaks: Bands or Lines (nm)	Resol. (Å)	T_{sol} (°C)	Liq P_{vap} (Torr)	Ref.
C_2Cl_4	Ar	20	C_2^* ($d^3\Pi_g - a^3\Pi_u$), weak continuum		−16	1.3	171
	Ar	20	C_2^*, strong continuum: Cl_2^* ($A^3\Pi° - X^1\Sigma^+$)		0	4.1	153
CCl_4	Ar	20	Broad continuum: Cl_2^* ($A^3\Pi° - X^1\Sigma^+$)		−7	23	153
Nitroethane	Ar	20	NO^* ($B^2\Pi - X^2\Pi$), cont.		−19	1.3	171
	Ar	20	Broad continuum, max ~420		−9	2.4	153
Dodecane	Ar	20	C_2^* ($d^3\Pi_g - a^3\Pi_u$)		−4	0.006	171
	Ar	20	C_2^* (564, 517, 474, 438), CH (431), strong		4	0.012	153
	X/Y nX/nY		cont: C_2H^* or liq. phase emis. of C_2^* or ...				
	Ar/N_2 85/15	20	C_2^*, CN^*, medium continuum		4	0.012	153

(continued)

Table 4B. Continued

Liquid	Gas	Freq. (kHz)	Peaks: Bands or Lines (nm)	Resol. (Å)	T_{sol} (°C)	Liq P_{vap} (Torr)	Ref.
	O_2	20	broad cont: CO_2^* quenching of C_2^* by O_2,		4	0.012	153
	Ar/O_2 90/10	20	CH* (391, 431), OH* (310–315), C_2^*, strong cont: CO_2^*		4	0.012	153
Dodecane+1,2-diaminoethane	Ar	20	CN*, C_2^*, strong continuum		4		153
Dodecane+NH_3	Ar	20	CN*, C_2^*	20			153
Dodec+Cr(CO)$_6$	Ar	20	Cr* (unresolved ~360, ~430)	20	16		172
Dodec+Fe(CO)$_5$	Ar	20	Fe* (unresolved 360, 380, 390, 410, 440)	20	5		172
1-Octanol	Ar	20	C_2^*, CH*	20	12	0.015	153, 172
			I_{rel}				
1-octanol + KOC$_8$H$_{17}$	Ar	20	K* (redshift ~10 Å, broad. ~40 Å) 170	~2			145
	Ar/He 50/50	20	K* (redshift ~10 Å, broad. ~40 Å) 120	~2			145
	Ar/He 20/80	20	K* (redshift ~10 Å, broad. ~40 Å) 40	~2			145
	He	20	K* (redshift ~10 Å, broad. ~40 Å) 10	~2			145
KOC$_8$H$_{17}$ in 1-octanol	Ar	20	K* (768, 771) redshift ~10 Å, broad. ~45 Å	~2		0.054	145
1-pentanol	Ar	20	Idem lower intensity			0.82	145
1-butanol/1-octanol	Ar 25/75	20	Idem lower intensity			0.95	145
Si oil Dow200	Ar	20	C_2^* with rovibronic struct., CH*	~1	4	~0.01	153
	Ar	20	C_2^*, rovibronic struct. calc $\Rightarrow$ 5000 K	4	0		173
Si oil + Fe(CO)$_5$	Ar	20	Fe* (372, 374, 375, 376, 382, 386), no shift	4	70		172
Si oil + Mo(CO)$_6$	Ar	20	Mo* (380, 386, 390), no shift	4	70		172
Si oil + W(CO)$_6$	Ar	20	W* (401, 407, 430), no shift	4	80		172
Si oil + Cr(CO)$_6$	Ar	20	Cr* (425, 427.5, 429), no shift	4	70		172
Si oil + Cr(CO)$_6$	Ar	20	Cr* line broadening $\Rightarrow$ 1700 atm	0.36	70		174

(continued)

85

Table 4B. Continued

Liquid	Gas		Freq. (kHz)	Peaks: Bands or Lines (nm)		Resol. (Å)	T_{sol} (°C)	Liq P_{vap} (Torr)	Ref.
				I_{rel}	Gas Therm Cond				
Si oil + Cr(CO)$_6$	Ar/He	50/50	20	Unresolv. Cr*: ~1.3	201 x 10^{-6} cal/s cm K	low	70		175
	Ar/Ne	50/50	20	Unresolv. Cr*: ~2.0	79 x 10^{-6}	low	70		175
	Ar		20	Unresolv. Cr*: ~3.0	43 x 10^{-6}	low	70		175
	Ar/Kr	50/50	20	Unresolv. Cr*: ~3.5	33 x 10^{-6}	low	70		175
	Ar/Xe	50/50	20	Unresolv. Cr*: ~5.8	27 x 10^{-6}	low	70		175
				I_{rel}	C_p / C_v				
	Ar		20	Unresolv. Cr*: ~3.7	1.67	low	70		175
	Ar/N$_2$	50/50	20	Unresolv. Cr*: ~0.6	1.50	low	70		175
	Ar/CO	50/50	20	Unresolv. Cr*: ~0.6	1.50	low	70		175
	Ar/CF$_4$	50/50	20	Unresolv. Cr*: ~0	1.22	low	70		175
		99/1	20	Unresolv. Cr*: ~0	1.63	low	70		175
	Ar/C$_2$F$_6$	50/50	20	Unresolv. Cr*: ~1.6	1.13	low	70		175
		99/1	20	Unresolv. Cr*: ~0.8	1.61	low	70		175

Table 4C. SBSL Spectra

Liquid	Gas	Freq. (kHz)		Spectral Data	T_{sol} (°C)	1300 nm (W/nm)	Ref.
Water	Air	27		Continuum (200–700 nm) (max ~235 nm)	22	4 10^{-11}	135
				Continuum (200–700 nm) more intense in UV	10	4.5 10^{-11}	135
			% glyc				
Water	Air	43	0%	Continuum (180–750 nm) (max ~220–240 nm)		4.5 10^{-11}	176
Water + glycerin	Air	43	10%	Continuum (400–650 nm)			176
		46	25%	Continuum (180–750 nm)		3.5 10^{-11}	176

(continued)

Table 4C. Continued

Liquid	Gas	P	Freq. (kHz)	Spectral Data	T_{sol} (°C)	$I_{300\ nm}$ (W/nm)	Ref.
Water		40%	52	Continuum (200–550 nm)		$2\ 10^{-12}$	176
	He	P Torr					
		150		Continuum (200–800 nm) max <200 nm	0	$3\ 10^{-10}$	28
		150		Idem	5	$2\ 10^{-10}$	28
		150		Idem	10	$1\ 10^{-10}$	28
		150		Idem	15	$6\ 10^{-11}$	28
		150		Idem	20	$4\ 10^{-11}$	28
		150		Idem	25	$2\ 10^{-11}$	28
Water	He, ³He	3		Continuum (200–800 nm) max ≤200 nm	0	$3\ 10^{-10}$	28
		3		Idem	25	$3\text{–}4\ 10^{-11}$	28
	Ne	3		Continuum (200–800 nm) max ≤200 nm	0	$3\ 10^{-10}$	28
		3		Idem	25	$5\ 10^{-11}$	28
	Ar	3		Continuum (200–800 nm) max ≤200 nm	0	$1\ 10^{-9}$	28
		3		Idem max ~ 210–240 nm	25	$1\ 10^{-10}$	28
	Kr	3		Continuum (200–800 nm) max ~ 220–240 nm	0	$2\ 10^{-9}$	28
		3		Idem max ~ 240–340 nm	25	$1.4\ 10^{-10}$	28
	Xe	3		Continuum (200–800 nm) max ~ 240–340 nm	0	$2\ 10^{-9}$	28
		3		Idem max ~ 260–380 nm	25	$1.6\ 10^{-10}$	28
Water	He	150	33	Continuum (200–700 nm) max <200 nm	24	$1.8\ 10^{-11}$	146, 28
	Ar	150	33	Continuum (200–600 nm) max ≤200 nm	24	$1.3\ 10^{-11}$	146, 28
	Xe	150	33	Continuum (200–700 nm) max ~ 240–300 nm	24	$4.5\ 10^{-11}$	146, 28
	X/Y	nX/nY					
	N₂/He	100/1	33	Continuum (200–700 nm) ≠ slope than He spectra	24	$4\ 10^{-11}$	146, 28

(continued)

87

Table 4C. Continued

Liquid	Gas	P Torr	Freq. (kHz)	Spectral Data	Resol.	T_{sol} (°C)	$I_{300 nm}$ (W/nm)	Ref.
	N_2/Ar	100/1	33	Similar to Ar spectra but 3× more intense		24	$4.6\ 10^{-11}$	146, 28
	N_2/Xe	100/2	33	Continuum (200–700 nm) max ~290 nm		24	$1.2\ 10^{-10}$	146, 28
	N_2/Xe	100/2		Bright: same shape as 24 °C spectra		1	$2.2\ 10^{-9}$	28
				Dim: slight increase of continuum in the UV		1	$2.3\ 10^{-10}$	28
Water	Ethane	2	24	Continuum (200–700 nm) no bands	10 nm	20	$2.5\ 10^{-11}$	147, 28
	Ethane	2	24	Continuum (200–700 nm) no bands	10 nm	0	$1.5\ 10^{-10}$	28
Ethanol	Xe	325	24	Continuum (280–700 nm) no bands	10 nm	–12	$1\ 10^{-11}$	147, 28
1-Butanol	Xe	70	24	Continuum (285–650 nm) no bands	10 nm	–8	$3\ 10^{-12}$	28
n-Dodecane	Xe	150	24	Continuum (300–700 nm) no swan bands	3 nm	–2	cutoff	147, 28
Water + NaCl (0.1 M)	air		32	Continuum (290–700 nm) no Na line	2 nm	25		165
+ KCl (0.1 M)	air		32	Continuum no K line	2 nm	25		165
H_2O	^{3}He	3	42	Continuum (200–700 nm) max ~ 220 nm		0	$3.5\ 10^{-10}$	143
	^{4}He	3	42	Very similar		0	$2.8\ 10^{-10}$	143
D_2O	^{3}He	3	40	Continuum (200–700 nm) max 220–260 nm		3	$5.6\ 10^{-11}$	143
	^{4}He	3	40	Very similar		3	$7.7\ 10^{-11}$	143
H_2O	H_2	3	42	Continuum (200–700 nm) max 230–290 nm		0	$2.5\ 10^{-11}$	143, 144
	D_2	3	42	Very similar		0	$3.2\ 10^{-11}$	143, 144
D_2O	H_2	3	40	Continuum (200–700 nm) max 450–500 nm		3	$6.6\ 10^{-12}$	143, 144
	D_2	3	40	Very similar		3	$7.5\ 10^{-12}$	143, 144
H_2O	Xe			Cont. max 280–360 nm, spher. reactor			$2.6\ 10^{-10}$	144
D_2O	Xe			Batch 1: cont. max 340–360 nm, spher. reactor			$1.8\ 10^{-10}$	144
	Xe			Batch 2: cont. max 320–360 nm, cyl. reactor			$3\ 10^{-10}$	144
	Xe			Batch 3: cont. max 290–340 nm, cyl. reactor			$6.4\ 10^{-10}$	144
H_2O + 1-butanol (2.5 ppm)	Xe			Impurities (ppm) affect intensity & spectral density			$3.5\ 10^{-9}$	144

Table 4D. SL from a Single Bubble Near a Surface

Liquid	Gas	Gas P (Torr)	Freq. (kHz)	Spectral Data	T_{sol} (°C)	$I_{300\,nm}$ (W/nm)	Ref.
Water (SB for ref)	Xe/O$_2$	1.5/150	30	Continuum (200–800 nm) (10-nm resol.)	12	$1.3\ 10^{-10}$	155
(SB near surf)	Xe	300	30	Cont. (200–800 nm) with upturn to the red very similar to SBSL spectrum (10-nm resol.)	12	$6.4\ 10^{-9}$	155
						$I_{400\,nm}$ (W/nm)	
n-Dodecane	Xe	290	30	Cont. (360–680 nm) + swan bands (10-nm resol.)		2–$3\ 10^{-11}$	155
Water + NaCl	Xe			No Na lines, no OH* band			155
Water + hypochlorite	Xe			No evidence of excimer (Xe–Cl) activity			155

4. SINGLE-BUBBLE SONOCHEMISTRY AND SONOLUMINESCENCE

4.1 Are Sonochemistry and SL Coupled?

One of the fundamental points is to determine whether the mechanical, chemical, and luminescent activities of bubbles are correlated. In fact, this question may be broken down as follows: (1) Are the sites of these activities different? (in case of acceptance, is there a common mechanism?) (2) Are the periods over which these activities develop different? (3) Is it possible to observe experimentally one activity in isolation from any other? Questions (1) and (2) will be broached in Section 5.3. In the present section, we will focus on the possible correlation between sonochemical and SL activity. In the past, Flynn [23] studied this correlation in multibubble fields and arrived at the conclusion that the relatively good parallelism between sonochemical and SL activity was indicative of the common origin of multibubble sonochemistry and SL. However, these systems are extremely complicated because bubbles of different sizes are (probably) loosely coupled and their growth and collapse somewhat random, so that the possibility cannot be rejected that chemical (but not luminescent) bubbles coexist with luminescent (but not chemically active) cavities. It is difficult to say whether sonochemistry and SL do or do not result from the same mechanism.

Recently, Lepoint et al. [107] investigated the chemical and SL activity of a single air bubble floating in a solution of the Weissler reagent (i.e., an aqueous NaI solution containing CCl_4 and starch [116]) (24- or 43-kHz acoustic setup). The main advantage of the "single bubble" system is that the different dynamics under which SBSL occurs can be easily monitored (Section 3.1.2). The chemical sequence is that Cl· radicals resulting from the intracavity breakdown of CCl_4 oxidize I^- ions, so that molecular iodine is produced in solution. Iodine molecules and starch form a blue complex detectable with the naked eye. It must be noted that Weissler's system is not a unique case. Indeed, as mentioned in Section 3.1.4.2, single air bubbles oscillating in a CS_2 aqueous solution transform carbon disulfide into colloidal sulfur [84] with an easily observable sulfur cloud. However, we will focus on the Weissler reaction.

In the latter case, chemical activity is triggered off for the four dynamic regimes, dancing, shuttlecock (nonsonoluminescing), stable (nonsonoluminescing), and stable sonoluminescing (see Section 3.1.2 for the definitions). In fact, a single weak blue thread indicating the complexation of iodine by starch originates from the bubble (Figure 57). This filament formed in a direction (upward and downward) that is likely to change during the experiment. The origin of this change in direction remained elusive until now. However, visual observations led us to suspect that dust trapped on the bubble surface could play a role. The time taken for the formation of the filament over a few centimeters is about 1 min.

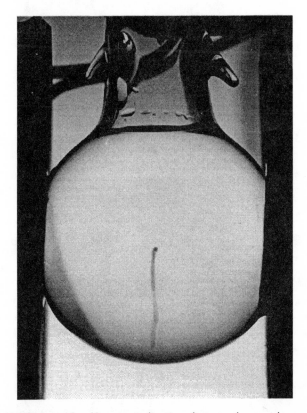

Figure 57. Formation of a filament indicating that starch complexes with iodine produced by the oxidation of I^- by $Cl\cdot$ radicals formed by the intracavity breakdown of CCl_4. The bubble (the shuttlecock) is at the center of the 30-mL reactor (ν_{ac}: 43 kHz). A typical solution consists of 200 mL of 1 M NaI and 10 mL of an aqueous solution of starch (4 g/L) to which 20 mL of a saturated aqueous solution of CCl_4 (kept under vacuum) has been added. (After Lepoint et al. [107].)

Dosimetry reveals that the "shuttlecock" bubble (which is nonsonoluminescing) is chemically speaking the most efficient (Figure 58; experimental conditions: see Figure 57). It is consequently clear that the intensity regimes of strong chemical activity and the strong SL of a single bubble are clearly separated.

More significant is the fact that a stable *nonluminescing* bubble is chemically active, even though only weakly (in this particular case, we have not yet succeeded in quantifying the amount of iodine). This regime is often described as "bouncing" in reference to the large-amplitude radii occurring after collapse (Figure 31b). The term *nonluminescing* means that a PMT able to detect 3×10^{-11} lm was unable to record any light from such a bubble (this latter information has been confirmed independently by other investigators [28,113]).

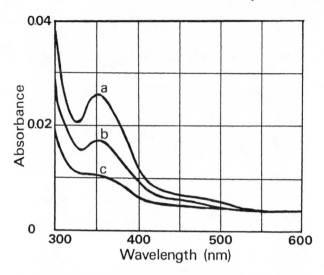

Figure 58. Comparison of the absorption spectra after 2 h of insonation in the presence of starch (see Figure 57 for the experimental details): (a) shuttlecock regime; (b) sonoluminescent regime [number of photons in the visible wavelength range (cf. Section 3.3.1): $(4 \pm 2) \times 10^3$]; (c) reference solution (insonation without bubble). (After Lepoint et al. [107].)

A partial conclusion (developed further in Section 5.2) is that the observations by Lepoint et al. [107] suggest that (1) the conditions giving rise to sonochemistry and SL are different in either type or intensity, (2) Flynn's conclusions (see the beginning of the present section) should not be transposed to single bubbles, and (3) one of the conclusions of the theoretical predictions by Yasui [65] (i.e., "Numerical calculations also reveal that no chemical reactions take place under a condition of non-light emission") is not relevant (Section 2.7.3).

4.2 Dosimetry: How Many Radicals Are Produced per Collapse?

As soon as a single isolated bubble is chemically active the matter arises as to measurement of the number of radicals produced per acoustic cycle. It must be emphasized that this information cannot be obtained in a multibubble field since the number of active bubbles cannot be determined.

(1) Using both the data reported in Figure 58 and an appropriate calibration curve (absorbance at 350 nm) and (2) assuming no side reactions, Lepoint et al. [107] deduced that the number of Cl atoms released in the solution per acoustic cycle was $\sim 10^8$ in the shuttlecock regime and $\sim 4 \times 10^7$ under SBSL conditions.

The experiments with single-bubble sonochemistry were essentially conclusive on the following points:

1. Single bubbles are chemically active and the phenomenon is not restricted to the Weissler reaction (see Section 3.1.4.2),
2. Since single stable nonluminescent bubbles are chemically active (though only weakly), it may be concluded that sonochemistry and SL are not correlated (at least in the case of single levitating bubbles). The difference between the number of photons ($\sim 10^4$ to 10^5 in the UV–visible) and chlorine atoms ($\sim 10^8$) in one acoustic cycle tends to confirm this conclusion.
3. The conditions giving rise to sonochemistry and SL are different in either type or intensity (see Section 5.1).

5. THE PROBLEM OF THE ORIGIN OF SL

5.1 Constraints Associated with Time

5.1.1 SBSL

The present analysis is associated with Sections 3.2.1.2 and 3.2.1.3. In the latter we mention that it was clearly demonstrated that the SBSL flash occurred near maximum collapse. Since this corresponded to a case in which the intracavity medium was strongly compressed and heated, the first reaction was to exploit the hot spot model (Section 2.7). As shown by Kamath et al. [63] and Kamath (see [177]), who used the "complete model" (see Section 2.7.3), intracavity temperatures due to compression could be more than 2000 K for about 20 ns under typical SBSL conditions. In comparison, the mean duration of an SL flash is ~ 100 ps. This led Crum [177] to ask: How can the bubble content remain compressed for so long without radiating?

In fact, it can be envisaged that radiative recombinations of radicals produced via thermochemical reactions [27] (triggered off during collapse and part of the rebound) are at the root of SBSL. However, in this case, it must be acknowledged that the temporal development of SBSL must be similar to the temporal development of the intracavity temperature (see Figure 15). This time (~ 10 ns) is too great with respect to the duration of the SL flash.

It could be argued that SBSL arises from the radiation of a blackbody since SBSL spectra can be fitted by means of Planck's law (the equivalent temperature is $\sim 20,000$ K). However, the work by Gompf et al. [120] (Section 3.2.1.2) affords a supplementary constraint. In fact, their experiment showed that the durations of the SL pulses for "red" or "blue" photons are the same. If the radiation came from progressive heating (due to bubble collapse) followed by the progressive cooling (the expansion of the bubble during the first rebound) of a blackbody radiator, red photons would be emitted for longer periods than blue photons. This is ruled out by experiment.

Although it must be borne in mind that SL emissions are triggered off near maximum bubble collapse (Section 3.2.1.3), the measurement of the flash duration in the SBSL mode enables a hot spot mechanism for the excitation of the intracavity medium to be discarded.

5.1.2 MBSL

Observations on MBSL are often explained in the context of the hot spot model [24,27]. However, Matula et al. [131] seriously question this model in view of the fact that refined measurements reveal that the flash duration for multibubble fields is "much less than 1 ns" (Section 3.2.2.2). These authors wrote: "It is interesting to note that our results deviate from that predicted by Kamath et al. [63], a state-of-the-art model incorporating well-accepted theories for bubble dynamics in the absence of shock-wave formation. Though the model assumes low applied pressures, relative to the conditions of our experiments, we would expect that under higher applied pressures, the temperature profiles would change such that the peak temperature would increase, and the 'width' of the profile would, at the very least, not decrease. Since our observations indicate a much shorter pulse width than predicted in the model [NB: see Section 2.7.3], this suggests that a fundamental mechanism occurring within sonoluminescing bubbles is not accounted for in present models that incorporate Rayleigh–Plesset-type dynamics [NB: see Sections 2.4 and 2.7], and this mechanism is the dominant mechanism for the observed light emission." They also say: "Our results suggest that sonoluminescence from cavitation fields is similar to that from single bubbles, in the sense that both types of SL have extremely short pulse durations." Section 3.4.3 can also be considered with due regard to this latter remark.

Though the temporal constraints discard the possibility of a sonoluminescent (intracavity) hot spot in SBSL and in one case (at least) of MBSL, the situation is different in sonochemistry.

5.2   Constraints Associated with the Absence of Correlation between SBSC and SBSL

As soon as SBSL and one case (at least) of MBSL fail to be explicable in terms of the radiation of an adiabatically compressed and heated medium, the question arises as to whether the hot spot model has in fact a real predictive character in sonochemistry. In multibubble sonochemistry, ample evidence is at hand but since this topic is beyond the scope of this introduction to SL, the reader may like to refer to the modern reviews by Suslick [19] and Henglein [178].

Single-bubble sonochemistry (Sections 4.1 and 3.1.4.2) also provides evidence in favor of the formation of a sonochemical hot spot via the compression of the intracavity medium. For example, in the Weissler reaction Lepoint et al. [107] demonstrate that if volatile and hydrophobic CCl_4 is replaced by a hydrophilic chlorinated compound (i.e., chloral hydrate in the same concentration as CCl_4), the formation of iodine cannot be detected. This emphasizes that the determining step (the cleavage of a C–Cl bond) occurs (mainly) inside the cavity in collapse.

Moreover, a "bouncing non-SL bubble" is shown to be less chemically active than an SL bubble (Section 4.1). This can be explained by the fact that the former's collapse is much less violent than the latter's (Section 3.1), so that it can be expected that the intracavity temperature and pressure increase will be less in the former case than in the latter. In this sense the hot spot model is predictive, qualitatively at least.

The picture so obtained is that sonochemistry is strongly associated with the increase in temperature and pressure inside a bubble in collapse (the formation of a hot spot) while SL results from a different mechanism which remains to be established.

5.3   Constraints Associated with Spectral Distributions: The Role of the Gas and the Liquid

5.3.1   What Can Be Done to Analyze SL Spectra?

From a general point of view, the analysis of emission spectra must be assisted by the qualitative knowledge of the source producing this light. In SL, this information is precisely what is sought after.

In order to illustrate the actual nature of the problem we can consider the case of an emission continuum which can be fitted by an 11,000 K blackbody, for example. By itself, the information contained in such a spectrum is not constraining enough for an analysis of the origin of the light to be unambiguous. For example, such a blackbody spectrum may correspond to (1) the chemiluminescence of a lighting worm [27], (2) the triboluminescence or the photoluminescence of an organic or inorganic crystal [93], or (3) the radiative recombination [$A^+ + e^- \rightarrow A + h\nu$ (continuum)] of a classic correlated plasma such as that generated either by shock waves traveling across a highly compressed gas or by the explosion of metal wires submitted to the passage of an intense electrical current [154]. Consequently, before any analysis is carried out, additional qualitative information on the nature of the emitting system must be obtained. This information is not available at the present time in SL. Moreover it may well be irrelevant to rely on the plausibility of a triggering mechanism. By way of an example, since SL is triggered off near maximum compression it would be logical to expect SL to arise from the radiation of the heated intracavity medium (the hot spot model, see Sections 2.3 and 2.7). The fact that the experimental spectrum of a single air bubble in water can be fitted by the spectral distribution of a 20,000 K blackbody could be a confirmation of a hot spot mechanism (though this temperature is somewhat high). However, although this analysis is logical, it is not sufficiently constrained. Because it is very short and consequently incompatible with a hot spot mechanism, the SL flash duration provides a determining constraint (Section 5.1).

If there is a lack of sufficient constraint (which is the case at the time of preparation of the present text), the emission spectra will not reveal all of the information that they contain. However, there is nothing against a given spectrum being analyzed within the context of a method of light diagnosis. Nevertheless, it must be borne in mind that such an approach will only enable the plausibility of the underlying model to be checked. An approach of this type has been carried out by Lepoint-Mullie et al. [179], who analyzed the spectral distribution of a single Ar bubble in water by optical plasma diagnostics, i.e., a method based on quantum and statistical mechanics. The essential conclusion was that it is plausible for SL to arise in a dense (i.e., nonideal) plasma [154] likely to have crystalline properties (without metallic behavior). As has been indicated above, thus far, there have been indications, but no definitive proof, that the origins of SL may involve a plasma.

The situation in SL is still more tricky. The most popular view is that SL arises (mainly if not exclusively) in the interior of bubbles. The experiment by Bourne and Field [75] (see Section 2.8.1) provides support for this view and also for the one that the impact of a liquid jet on the opposite wall of a bubble is at the root of SL. Since the bubbles studied by Bourne and Field had the peculiarity of being large (initial radius ~6 mm) and cylindrical, the question arises as to whether or not acoustically driven cavitation bubbles behave similarly. (The detection of photons associated with the compression of the lobes may possibly be due to the greater amount of intracavity gas in such bubbles when compared with acoustic bubbles). As far as the present authors are aware, the problem of the site of SL for acoustically driven bubbles has never received a definite answer.

5.3.2 Is SL Related to Vacuum Quantum Radiation?

In Section 2.9, we mentioned several references related to different models. Among them references [94] and [95] are connected with the possibility that SL is due to the emission of photons from vacuum. In terms of quantum electrodynamics this emission is known as the Casimir effect [180]. Casimir energy is associated with the force (of quantum origin) causing two planes of atoms to undergo mutual attraction. Schwinger [94 and internal references] demonstrated the general character of the Casimir energy and suggested that the oscillation of the wall of an acoustic bubble could trigger off the release of Casimir energy. This suggestion created a sensation since the phenomenon of vacuum quantum radiation (though predicted by the laws of quantum electrodynamics) has not yet been observed. This stimulated the work by Eberlein [95].

Several theoretical counterarguments have now been reported [96]. However, the best evidence against the model of vacuum quantum radiation for SL is due to the fact that the dissolved gas plays a major role in SL intensity and spectral distribution while this should not be the case in Casimir energy model of SL, as emphasized by Unnikrishnan and Mukhopadhyay [181].

5.4 The Origin of SL: The Current Situation

In spite of more than 15 explanatory models reported in the literature (see Section 2.9), the mechanism of light emission remains elusive. Three conditions for a model to be accepted are that it must (1) obey the laws of fluid mechanics, (2) obey the laws associated with the properties of light, and (3) describe all of the experimental features. At the present time, practically no model explains *all* of the experimental features. Because of the extreme complexity, a discussion about this aspect is beyond the scope of this introduction.

Consequently, we will confine ourselves to report the picture given by the literature. Since only a few models obey rigorously the laws of fluid dynamics, two explanatory descriptions emerged. At the present time the converging shock or pressure wave model describing SL as the light emission from a plasma (or energized gas) is the most popular (see Sections 2.7.4 and 2.9). However, Prosperetti's model appears to be a possible challenger (see Section 2.8.3).

However, we must acknowledge that a general consensus has not yet been attained, in particular, because the versatile role of small amounts of solutes, the effect of the temperature of the host liquid, and so forth remain unexplained in relation to the current models.

5.5 A Compulsory Precondition and an Unanswered Question: What Is (or Are) the Site(s) of SL Activity?

Three main effects—acoustic emission, sonochemical effects, and SL—are associated with bubble collapse. The site where acoustic emissions take place has been known since the work by Lord Rayleigh [33]. During collapse, the pressure in the liquid immediately adjacent to the bubble–liquid interface dramatically increases. Refining Rayleigh's analysis (Section 2.3), Hickling and Plesset [182] studied the case of an incompressible liquid. For a bubble expanding to 10 times its initial radius (this corresponds to the typical case of a single luminescing bubble), calculations [182] show that the collapse is accompanied by an increase in pressure in the vicinity of the bubble (~20,000 atm for a few nanoseconds over a distance of a few times the minimum bubble radius). This overpressure translates into a diverging shock wave which is rapidly transformed into a sound wave detectable by a hydrophone. The emission from a single bubble in levitation has been observed by Matula et al. [183].

As showed by Suslick's group, there are two reaction sites in sonochemistry, namely, the intracavity gaseous phase and part of the liquid surrounding the bubble [19]. In SBSL and MBSL the situation is not so clear.

We can first consider SBSL and begin with the theoretical aspect. Prosperetti [83] (see Section 2.8.3) demonstrates that in a resonant acoustic field the collapse of an initially spherical and isolated bubble must end with an inward jet violently striking the opposite bubble wall (this is due to the coupling of radial and translational motion of a bubble; see also Longuet-Higgins [82]). Prosperetti assumes that the next step could for the impact of the jet to be accompanied by a fracture. This point of view is based on the fact that the jet impact is so rapid that the molecules in the jet are "frozen," so that the jet can be considered as a solid likely to undergo fracture. SL seems to be fractoluminescence [93a,b].

Second, as far as the experimental aspect is concerned, Ohl et al. [184] recently studied the luminescence emitted by laser-generated bubbles and noted that SL arises in spherical and aspherical bubbles. Such laser-generated bubbles can be formed at varying distances from a solid wall. Their initial shape is spherical, but more or less significant distortions in shape occur during the collapse. An inward jet may develop because of the asymmetry in the liquid flow around the bubble, with this asymmetry being due to the presence of the solid boundary (Section 2.8.1; see also [29]). The development of the jet is directed toward the rigid boundary (Figure 19). The authors mention [184]: "The cavitation luminescence is visible as a bright spot at the center of the lower bubble wall . . . . The light emission occurs at the very late stage of bubble collapse and is a single event."

Two remarks must be considered: (1) it appears that the more rapidly the distortion occurs, the less intense the SL will be; (2) even if aspherical bubbles luminesce at the site of jet impact inside a bubble, the experiment will not be able to ascertain whether SL emission arises at the instant of jet impact or at the instant when the bubble attains minimum size. Moreover, it must be acknowledged that for reasons of conservation of momentum a bubble goes toward a boundary during its collapse (downwards in the case of Ohl et al.'s experiment). Consequently, the light spot observed by Ohl et al. [184] could be due to a bubble which moved downward without a jet necessarily being responsible for this light emisson.

The two points discussed above show that the question of the site of SL activity is actually a matter of debate.

The same problem has existed for MBSL since the hot spot model has been questioned (see Sections 3.2.2.2 and 5.1.2). We can consider two cases for which the site of SL activity is thought to be the liquid phase. The first experiment concerns the emission of alkali metals. Flint and Suslick [145] analyzed the position and shape of the K doublet ($^2P_{3/2} - {}^2S_{1/2}$ and $^2P_{1/2} - {}^2S_{1/2}$ transitions) for potassium 1-octanolate dissolved in various primary alcohols, and for different dissolved gases (Ar versus Ar/He mixtures). If it is assumed that potassium luminesces from the intracavity gas phase, it may be expected that a change in the vapor pressure of the liquid will induce a modification in the conditions responsible for (1) the transfer of an electron and (2) the excitation of the metal atom. However, the spectral width and position of the K doublet are independent of the vapor pressure of the cavitating liquid. Moreover, any excited species is sensitive to its surroundings. One of the most obvious disturbances is the Stark effect which is due to local electrical fields created by ions or electrons surrounding a light emitter [88]. As soon as the pressure increases, neutral species may also influence the light emitters during collisions because of van der Waals interactions [88]. Not only the line width changes with the pressure, but also the frequency of the emission. Indeed an emitter of light interacting with surrounding foreign atoms can no longer be considered as an isolated emitter. The shift associated with this effect depends on the nature of the atoms surrounding the emitter. As Chen and Takeo [185] showed, the alkali metals' doublet (among other systems) can be shifted toward the red zone of a spectrum if the disturbances are created by Ar atoms, and the blue zone when He is involved. Flint and Suslick did not record a modification in the shift of the K doublet for an increasing concentration of He in comparison with Ar as the dissolved gas. This led these authors [145] to hypothesize that alkali metals emit from the liquid phase.

The second system was studied by Lewshin and Rschevkin [186] a few years after the discovery of the SL phenomenon. These authors tried to determine whether SL arose from the gaseous or the liquid phase. Their reasoning was as follows: Let it be assumed that SL is due to chemiluminescent reactions in the cavitating liquid. Considering that pyrogallol is "a quencher of chemi-luminescence in the liquid phase" [186], these authors suggested adding pyrogallol to cavitating water. They did not observe any strong decrease in intensity and consequently concluded that SL arose from the gas phase.

First, it must be emphasized that there is no a priori reason to expect SL to be due to a chemiluminescent system (though it is indeed a possiblity). Second, Lewshin and Rschevkin do not specify the types of systems that pyrogallol is able to quench. Quenching is not a general process. In order to illustrate this assertion let us consider the case of $RuL_3^{2+} \cdots MW_{11}^{6-}$ ion pairs in aqueous solutions [187], where (1) RuL_3^{2+} is the fluorescent light emitter activated by an external light source (L is a ligand) and (2) the quencher is $Co(H_2O)SiW_{11}O_{39}^{6-}$. By an appropriate choice of complex RuL_3^{2+} it has been possible to tune the process of quenching of an electron transfer. If $Ru[bpy]_3^{2+}$ (with bpy = 2,2'-bipyridine) is used, quenching is efficient. However, if only one ligand is changed (i.e., $Ru[bpy]_2[biq]^{2+}$ with biq = 2,2'-biquinoline is used), quenching is totally suppressed. Since quenching is not a general process for a given moleucle, it can be deduced that the experiment by Lewshin and Rschevkin is not necessarily constraining.

The conclusion drawn from this section is that the site of MBSL and SBSL remains largely undetermined. The present authors think that this problem is still crucial not only because several SL models involving the liquid phase immediately adjacent to the liquid interface have been reported in the literature (Prosperetti [83], Lepoint et al. [64]), but also because the process of collapse [with the formation of enormous pressures in the liquid around a bubble (the beginning of this section)] could significantly modify the structure of the matter constituting the shell.

Though there is no trivial indication that such a structuring is associated with SL, it would probably be interesting to determine the physicochemical properties of the medium involved in this shell. Not only can the pressure in this region be $>10^4$ bars for a few nanoseconds, but the gas content is greater than the equilibrium value because of the outgoing flow due to the collapse (Section 2.6.1.2).

In conclusion, our objective is to emphasize that the incertitude on the nature of the site(s) of SL activity deserves a careful analysis of the question itself and its possible physicochemical consequences.

6. CONCLUSION

Three main effects—acoustic emissions, sonochemical effects, and SL—are associated with bubble collapse. These effects were known for a long time in the case of bubbles belonging to multibubble clouds. However, the complexity of these systems makes any access to quantitative information difficult. Recently, acoustic emissions, sonochemical effects, and SL have been shown to accompany the collapse of single bubbles maintained in levitation. Since a single bubble is highly controllable, a great deal of quantitative information can be obtained.

From the combination of the elements obtained to date, the following picture emerges.

During the violent collapse of a bubble, a pressure pulse takes place in the liquid shell immediately adjacent to the bubble interface, frees itself from this interface, and travels in the liquid in the form of a diverging shock wave. While this occurs outside the bubble, the intracavity medium is compressed and heated. In typical violent conditions the period over which these events occur is 10–100 ns. The pulses of pressure just outside the bubble may reach 10^4 bars, while the peaks of temperature and pressure inside the cavity may reach 10^3–10^4 K and 10^3 bars, respectively. In the case of a single bubble and the particular conditions associated with Weissler's reaction, up to 10^8 Cl· radicals can be released per acoustic cycle. The theory of Lord Rayleigh (improved by Hickling and Plesset) gives a good description of the physical conditions occurring in the liquid shell around a bubble. The hot spot theory gives a semiquantitative description of the physical conditions (thermodynamical temperatures and pressures) occurring inside a bubble in collapse. Even if the quantification of the microscopic events triggered by these transient physical conditions is difficult and still lacking, the sonochemical hot spot is on solid ground.

The experiments on single-bubble sonochemistry demonstrate that chemical events (the production of radicals) and SL are not coupled. This reinforces the observations according to which SL (i.e., single- and multibubble SL) is due to a mechanism other than a hot spot, though this mechanism remains elusive.

SL exhibits remarkable characteristics of time, geometry, and energy. Concerning the time scale, the characteristic time is the acoustic period (~ 50 µs) while the SL flash duration is about 50 ps, i.e., a contraction factor of 10^6. The geometrical scale is contracted by about 10^4 since the characteristic length of the acoustic system is centimetric while the mean radius of a bubble is micrometric. Moreover the variation of volume for a typical single sonoluminescing bubble is as high as 10^6 if the volumes of a bubble at maximum expansion and collapse are compared. This variation is perhaps greater in MBSL. The energy concentration is striking: The energy density of the sound field is $\sim 10^{-10}$ eV per atom while the energy of emitted photons is of the order of the electron-volt, i.e., a concentration factor of 10^{12}. These data are directly connected with the case of a single bubble in levitation in a resonant acoustic setup. Is the situation transposable to MBSL? In fact, at the present time, the initial feeling that single- and multibubble SL are different (because of the differences in spectral distributions) tends partially to collapse. For example, the flash duration in both SBSL and one case (at least) of MBSL is extremely short and a single- and multibubble cloud are both chemically active. Moreover, in the case of dodecane as the host liquid, the emission spectrum of a single bubble near a solid boundary (a continuum with weak C_2^* bands) seems to be intermediate to the spectra of an ~20-kHz multibubble cloud (continuum with C_2^* bands) and a single bubble in levitation (a continuum, only). This tends to suggest that the fundamental triggering mechanism(s) of SL in various situations could be qualitatively the same but that the local conditions lead to a set of different results.

However, at the present time, in spite of a great deal of remarkable experimental and theoretical investigations, the origin of the phenomenon is still unknown (to date, 15 models have been proposed).

As for every physical, chemical, or biochemical luminescent system, the string of events giving rise to photons is complex. In SL, this complexity is increased by (1) the versatile and nonlinear character of the bubble behavior, (2) the fact that there are very many possible ways for the conversion of the mechanical energy associated with the bubble oscillation into light, and (3) the technical difficulty in obtaining very constraining experiments. Concerning point (1), we noted that several mechanical events can be invoked as triggering factors. Considering the literature, it appears that two of these factors are highly realistic from the point of view of fluid dynamics: a supersonic collapse and the formation of an inward jet. This situation reveals a change in the study of SL. It is now acknowledged that a bifurcation (in the physical sense) occurs in the bubble history. The converging shock or pressure wave model describing SL as a plasma or a gas charged with energy involves an "extraordinary" event (because it is superimposed on the compression of the intracavity medium) and leads to unexpected results. The formation of an inward jet accompanied (for example) by the fracture of the latter (Prosperetti's model of fractoluminescence) also appears as an "extraordinary" event to which a bifurcation in the observable results occurs with bubble life.

At the present time it may be expected that in addition to the current numerous theoretical suggestions relating to SL in the literature, others will be formulated in the future. In view of the high degree of activity in this field, we decided not to structure the present introduction around explanatory models of SL but rather to introduce the basic elements of the life of bubbles.

In the search for the establishment of a bridge between the opposite viewpoints of "mechanics" and "electromagnetic emission," there are two possibilities but these involve radically different approaches. One indicates that "the gas phase is activated," the other that "the liquid phase is involved." It should be borne in mind that even when the points of departure and arrival (our opposite viewpoints) have been determined, the bridge still remains to be built.

We believe that at the present time, the determination of the site(s) of SL is a keystone for a better understanding of the origin of this extremely fascinating and mysterious phenomenon. At the very least a definite answer to this problem should provide fundamental restrictions.

Be that as it may, the last word should fall to poetry:

> *When mysteries are intelligent, they hide in the light.*
> Jean Giono [Manosque 1895; Ibid. 1970]
> in *Ennemonde*

ACKNOWLEDGMENTS

We are very grateful to Professor T. J. Mason for his confidence. We would like to express our thanks to Professor A. Prosperetti who through his numerous and rigorous pieces of advice has contributed significantly to our fascination with the topics discussed in this chapter. It is to him also that we owe our thanks for a chapter on the theoretical aspects of sonochemistry which was published in a text entitled "Synthetic Organic Sonochemistry" edited by J-L. Luche and published by Plenum in 1998. We are also very indebted to Professors R. Avni, C. Deutsch, L. A. Crum, J. Frohly, M. and A. Goldman, A. Henglein, W. Lauterborn, M. S. Longuet-Higgins, R. A. Roy and Drs. M. Baus, G. Maynard and P. Supiot for numerous discussions over the last few years which have helped to expand and develop our ideas on the topics of acoustic cavitation, sonochemistry and sonoluminescence. We would like to thank the Institute Meurice, the Fonds National de la Recherche Scientifique (Belgium) and the Fondation A de Potter for financial support.

NOTES

1. Noltingk and Neppiras [39] considered that the "gas bubble/liquid" system responds in a quasi-static manner to the acoustic pressure traveling through the liquid. The term *quasi-static* means that if an observer was able to determine the position and velocity of each particle inside a bubble, this observer would not be able to detect whether the medium studied was changing or not. Such a situation is associated with transformations that are slow enough for dynamic effects to be neglected.

2. The acoustic pressure can be determined by three methods [29], i.e., (1) by means of a calibrated hydrophone, (2) calorimetrically, or (3) radiometrically. The first method requires the use of standard guidelines (defined in [41]) in order to avoid spatial averaging which occurs when the hydrophone is not small enough in relation to the acoustic wavelength. The two other methods give a temporal and spatial average for the acoustic intensity (I_{ac}). The link between I_{ac} and P_{ac} is $I_{ac} = P_{ac}^2/2\rho c$ (progressive waves). In this equation, c is the speed of sound in the liquid and ρ the specific mass (the product $\rho c = Z$, i.e., the impedance of the liquid). It must be noted that the radiometric and calorimetric methods cause departures from the actual local acoustic pressure. This problem is approached in [29].

3. Explicitly, Wu and Roberts [67] consider that adiabatic heating by itself cannot generate sonoluminescent emission.

4. Radiation through bremsstrahlung is due to the "braking" of a fast electron in the vicinity of an ion (electron–ion bremsstrahlung) or an atom (electron–atom bremsstrahlung). The emission of a photon from an electron in these conditions results from the fact that the electron only possesses a linear momentum. The presence of an ion or an atom is required for the energy conservation. The inverse process (i.e., inverse electron–ion or electron–atom bremsstrahlung) corresponds to the absorption of light by a slow electron in the vicinity of either an ion or an atom. Therefore, in inverse bremsstrahlung a slow electron becomes accelerated.

5. Models based on an electrical breakdown [85] have now been abandoned (for a review see [27]). In particular, the model of electrical intracavity discharges via bubble fragmentation [86] was recently shown to be incorrect [87].

6. Two systems of coordinates enable the description of the motions of a particle (in the sense of fluid dynamics), i.e., the Lagrangian and the Eulerian systems. In the Langrangian coordinates the observer is associated with the particles studied. In the Eulerian coordinates the observer is associated with an external referential. The use of one or the other system is essentially motivated by the simplicity in the establishment or the resolution of the problem.

7. In this context, the term *dipolar* has a loose sense. Actually the flow appears to be a "Stokeslet" [117]. This term is in close connection with the motion of a sphere with a uniform velocity in a viscous fluid. This movement is described by the Stokes equation, so that (1) the radial and tangential velocities of the surrounding fluid, (2) the drag coefficient, and (3) the field of pressure in the surrounding fluid may be easily determined. The case of a nonspherical body is much more complicated. However, for bodies with the symmetry of an even polyhedron, the properties quoted above are similar to those of a sphere. The corresponding equations for a sphere are modified by a coefficient depending exclusively on the geometry of the body considered. The field of velocity associated with such a case is a "Stokeslet." For example, the field of velocity induced at large distance by a small insect in steady motion in the air is a Stokeslet [117].

REFERENCES

[1] Marinesco, N., and Trillat, J.-J. *C. R. Acad. Sci.,* 196 (1933) 858.

[2] Frenzel, H., and Schultes, H. *Z. Phys. Chem. Abt. B*, 27 (1934) 421.

[3] Harvey, E. N. *J. Am. Chem. Soc.*, 61 (1957) 2392.

[4] Schmid, J. *Acustica*, 9 (1959) 321.

[5] (a) Jarman, P. D., and Taylor, K. J. *Br. J. Appl. Phys.*, 15 (1964) 321; (b) Ibid. 16 (1965) 675.

[6] Peterson, F. B., and Anderson, T. P. *Phys. Fluids*, 10 (1967) 874.

[7] Gimenez, G. *J. Acoust. Soc. Am.*, 71 (1982) 839.

[8] Buzukov, A. A., and Teslenko, V. S. *Sov. Phys. JETP*, 14 (1971) 189.

[9] Akmanov, A. G., Benkovskii, V. G., Golubnichii, P. I., Maslennikov, S. I., and Shemanin, V. G. *Sov. Phys. Acoust.*, 19 (1974) 417.

[10] Benkovskii, V. G., Golubnichii, P. I., and Maslennikov, S. I. *Sov. Phys. Acoust.*, 20 (1974) 14.

[11] Golubnichii, P. I., Dyadyushkin, P. I., Kalyuzhnyi, G. S., and Kudlenko, V. G. *J. Appl. Mech. Tech. Phys.*, 19 (1978) 472.

[12] Golubnichii, P. I., Dyadyushkin, P. I., Kalyuzhnyi, G. S., and Korchikov, S. D. *Sov. Phys. Tech. Phys.*, 24 (1979) 1006.

[13] Benkovskii, V. G., Golubnichii, P. I., Maslennikov, S. I., and Olzoev, K. F. *J. Appl. Spectrosc.*, 24 (1979) 964.

[14] (a) Golubnichii, P. I., Gromenko, V. M., and Filonenko, A. D. *Sov. Phys. Tech. Phys. Lett.*, 5 (1979) 233; (b) Ibid. *J. Appl. Spectrosc.*, 31 (1979) 1430; (c) Ibid. *Sov. Phys. Tech. Phys.*, 24 (1979) 1251; (d) Ibid. *Sov. Phys. Tech. Phys.*, 25 (1980) 1392.

[15] Belaeyva, T. V., Golubnichii, P. I., Dyadyushkin, P. I., and Lysikov, Y. I., *J. Appl. Mech. Tech. Phys.*, 22 (1982) 461.

[16] Leighton, T. G. *The Acoustic Bubble*. Academic Press, London, 1994.

[17] Margulis, M. A., and Grundel', L. M. *Dokl. Acad. Nauk SSSR*, 265 (1982) 914.

[18] (a) Lauterborn, W. *Frontiers in Physical Acoustics*. XCIII Corso, Soc. Italiana di Fisica, Bologna, 1986, pp.124–144; (b) Lauterborn, W. *Acustica*, 82 (1996) S46.

[19] Suslick, K. S. (ed.). *Ultrasound: Its Chemical, Physical and Biological Effects*. VCH, New York, 1988.

[20] Mason, T. J. (ed.). *Advances in Sonochemistry*. JAI Press, London, Vols. 1 (1990), 2 (1992), 3 (1993), 4 (1995).

[21] Gaitan, D. F., and Crum, L. A. *J. Acoust. Soc. Am.*, 87 (1990) S141.

[22] Gaitan, D. F., Crum, L. A., Church, C. C., and Roy, A. R. *J. Acoust. Soc. Am.*, 91 (1992) 3166.

[23] Flynn, H. G. In Mason, W. P. (ed.), *Physical Acoustics*, Vol. 1, Part B. Academic Press, New York, 1975, pp. 57–172.

[24] Walton, A. J., and Reynolds, G. T. *Adv. Phys.*, 33 (1984) 595.

[25] El'piner, I. E. (ed.). *Ultrasound, Physical, Chemical and Biological Effects.* Consultants Bureau, New York, 1964, Chap. 3.

[26] Margulis, M. A. *Sov. Phys. Acoust.*, 15 (1969) 135.

[27] Verrall, R. E. and Sehgal, C. M. in Ref. 19, Chap. 6.

[28] Barber, B. P., Hiller, R. A., Löfstedt, R., Putterman, S. J., and Weininger, K. R. *Phys. Rep.*, 281 (1997) 65.

[29] Lepoint, T., and Lepoint-Mullie, F. In Luche, J.-L. (ed.), *Synthetic Organic Sonochemistry.* Plenum, New York, 1998.

[30] (a) Neppiras, E. A. *Acoustic Cavitation.* North-Holland, Amsterdam, 1980; (b) Atchley, A. A., and Crum, L. A. in Ref. 19, pp. 1–64; (c) Blake, J. R., Boulton-Stone, J. M., and Thomas, N. H. (eds.). *Bubble Dynamics and Interface Phenomena.* IUTAM Symp. Proc., Birmingham, U.K. Kluwer Academic, Dordrecht, 1994; (d) Brennen, C. E. (ed.). *Cavitation and Bubble Dynamics.* Oxford University Press, New York, 1995; (e) Lauterborn, W. (ed.). *Cavitation and Inhomogeneities in Underwater Acoustics*, 1st International Conference. Springer-Verlag, Berlin, 1981; (f) Apfel, R. E. (ed.). *Methods in Experimental Physics*, Edmonds, P. D. (ed.). Academic Press, New York, 1981.

[31] Atchley, A. A., and Prosperetti, A. *J. Acoust. Soc. Am.*, 86 (1989) 1065.

[32] Thornycroft, J., and Barnaby, S. W. *Inst. C. Eng.*, 122 (1895) 51.

[33] Lord Rayleigh. *Philos. Mag.*, 34 (1917) 94.

[34] Kornfeld, M., and Suvorov, L. *J. Appl. Phys.*, 15 (1944) 495.

[35] Philipp, A., and Lauterborn, W. *J. Fluid Mech.*, 361 (1998) 75.

[36] Besant, X. *Hydrostatics and Hydrodynamics.* Cambridge University Press, London, 1859.

[37] Plesset, M. S. *J. Appl. Mech.*, 16 (1949) 277.

[38] Franc, J. P., Avellan, F., Belahadji, B., Billard, J.Y., Briançon-Marjolet, L., Fréchou, D., Fruman, D. H., Karimi, A., Kueny, J. L., and Michel, J. M. *La cavitation.* Presses Universitaires de Grenoble, Grenoble, 1995.

[39] (a) Noltingk, B. E., and Neppiras, E. A. *Proc. Phys. Soc. B*, 63 (1950) 674; (b) Neppiras, E. A., and Noltingk, B. E. *Proc. Phys. Soc. B*, 64 (1951) 1032.

[40] Poritsky, H. Proc. 1st U.S. National Congress in Applied Mathematics (ASME) 1952, p. 813. [The original analysis was carried out by Taylor, G. I.; see Taylor, G. I., and Davies, R. M. *The Scientific Papers of G. I. Taylor*, Vol. III, G. K. Taylor (ed.). Cambridge University Press, London, 1963, p. 337.]

[41] Harris, G. R. *Ultrasound Med. Biol.*, 11 (1985) 803.

[42] (a) Prosperetti, A., and Lezzi, A. *J. Fluid Mech.*, 168 (1986) 457; (b) Lezzi, A., and Prosperetti, A. *J. Fluid Mech.*, 185 (1987) 289.

[43] Flynn, H. G. in Ref. 23, p. 76.

[44] Herring, C. Office of Science Research and Development Report 236 (NDRC Report C-4-sr 10-010, Columbia University), 1941.

[45] Kirkwood, J. G., and Bethe, H. A. Office of Science Research and Development Report 588, 1942.

[46] Trilling, L. *J. Appl. Phys.*, 23 (1952) 14.

[47] Gilmore, F. R. Hydrodynamics Laboratory Report 26-4, California Institute of Technology, 1952.

[48] Keller, J. B., and Miksis, M. *J. Acoust. Soc. Am.*, 68 (1980) 628.

[49] Prosperetti, A., Crum, L. A., and Commander, K. W. *J. Acoust. Soc. Am.*, 83 (1988) 502.

[50] Church, C. C. *J. Acoust. Soc. Am.*, 97 (1995) 1510.

[51] Fujikawa, S., and Akamatsu, T. *J. Fluid Mech.*, 97 (1980) 481.

[52] Leighton, T. J. in Ref. 16, pp. 308–312.
[53] Leighton, T. J. in Ref. 16, p. 129.
[54] Minneart, M. *Philos. Mag.*, 16 (1933) 235.
[55] Harvey, E. N., Barnes, D. K., McElroy, W. D., Whiteley, A. H., Pease, D. C., and Cooper, K. W. *J. Cell. Comp. Physiol.*, 24 (1944) 1.
[56] (a) Blake, F. B. Technical Memo 12 Acoustic Research Laboratory, Harvard University, Cambridge, Mass., 1949; (b) Hsieh, D. Y., and Plesset, M. S. *J. Acoust. Soc. Am.*, 33 (1961) 206; (c) Eller, A. I., and Flynn, H. G. *J. Acoust. Soc. Am.*, 37 (1965) 493; (d) Safar, M. H. *J. Acoust. Soc. Am.*, 43 (1968) 1188; (e) Eller, A. I. *J. Acoust. Soc. Am.*, 46 (1969) 1246; (f) Skinner, L. A. *J. Acoust. Soc. Am.*, 51 (1972) 378; (g) Eller, A. I. *J. Acoustic. Soc. Am.*, 57 (1975) 1374; (h) Crum, L. A. *J. Acoust. Soc. Am.*, 68 (1980) 203; (i) Crum, L. A. *Ultrasonics*, 22 (1984) 215; (j) Strasberg, M. *J. Acoust. Soc. Am.*, 33 (1961) 359; (k) Gould, R. K. *J. Acoust. Soc. Am.*, 56 (1974) 1740.
[57] Cramer, E. in Ref. 30e, pp. 54–63.
[58] Crum, L. A., and Gaitan, D. F. *SPIE New Methods in Microscopy and Low Light Imaging*, 1161 (1989) 125.
[59] Flynn, H. G., and Church, C. C. *J. Acoust. Soc. Am.*, 76 (1984) 505.
[60] Hickling, R. *J. Acoust. Soc. Am.*, 35 (1963) 967.
[61] Young, F. R. *J. Acoust. Soc. Am.*, 60 (1976) 100.
[62] Nigmatulin, R. I., and Khabeev, N. S. *Fluid Dyn.*, 9 (1974) 759.
[63] Kamath, V., Prosperetti, A., and Egolfopoulos, F. N. *J. Acoust. Soc. Am.*, 94 (1993) 248.
[64] Lepoint, T., De Pauw, D., Lepoint-Mullie, F., Goldman, M., and Goldman, A. *J. Acoust. Soc. Am.*, 101 (1997) 2012.
[65] Yasui, K. *J. Phys. Soc. Japan*, 66 (1997) 2911.
[66] Wu, C. C., and Roberts, P. H. *Phys. Rev. Lett.*, 70 (1993) 3424.
[67] (a) Wu, C. C., and Roberts, P. H. *Proc. R. Soc. London Ser. A*, 445 (1994) 323; (b) Roberts, P. H., and Wu, C. C. *Phys. Lett. A*, 213 (1996) 59.
[68] Moss, W. C., Clarke, D. B., White, J. W., and Young, D. A. *Phys. Fluids*, 6 (1994) 2979.
[69] Moss, W. C., Clarke, D. B., White, J. W., and Young, D. A. *Phys. Lett. A*, 211 (1996) 69.
[70] Moss, W. C., Clarke, D. B., White, J. W., and Young, D. A. AIP Conf. Proc. (Pt. 1, Shock Compression of Condensed Matter, 1995), 1996, pp. 453–458.
[71] Landau, L., and Lifchitz, E. *Mécanique des fluides*. Editions MIR, Moscow, 1989.
[72] Coleman, A. J., Saunders, J. E., Crum, L. A., and Dyson, M. *Ultrasound Med. Biol.*, 13 (1987) 69.
[73] Lauterborn, W., and Bolle, H. *J. Fluid Mech.*, 72 (1975) 391.
[74] Lauterborn, W. in Ref. 30e, pp. 3–12.
[75] Bourne, N. K., and Field, J. E. *Proc. R. Soc. London Ser. A*, 435 (1991) 423.
[76] Jungnickel, K., and Vogel, A. in Ref. 30c, pp. 47–53.
[77] Oguz, H. N. in Ref. 30c, pp. 65–72.
[78] d'Agostino, L., and Brennen, C. E. ASME Cavitation and Polyphase Flow Forum, Houston, 1983, pp. 72–76.
[79] Chahine, G. L. in Ref. 30e, pp. 195–206.
[80] Plesset, M. S., and Prosperetti, A. *Annu. Rev. Fluid Mech.*, 9 (1977) 145.
[81] Frost, D., and Sturtevant, B. *ASME J. Heat Transfer*, 108 (1986) 418.
[82] (a) Longuet-Higgins, M. S. *Proc. R. Soc. London Ser. A*, 454 (1998) 725; (b) Ibid. *Proc. R. Soc. London Ser. A*, 453 (1997) 1551; (c) Ibid. *J. Acoust. Soc. Am.*, 100 (1996) 2678.
[83] Prosperetti, A. *J. Acoust. Soc. Am.*, 101 (1997) 2003.
[84] Verraes, T., Lepoint-Mullie, F., and Lepoint, T. with an Appendix by M. S. Longuet-Higgins. Submitted.
[85] Margulis, M. A. *Russ. J. Phys. Chem.*, 55 (1981) 81.
[86] (a) Margulis, M. A. *Russ. J. Phys. Chem.*, 59 (1985) 882; (b) Margulis, M. A. in Ref. 20 (Vol. 1), pp. 39–80.

[87]   Lepoint-Mullie, F., De Pauw, D., and Lepoint, T. *Ultrasonics Sonochem.*, 3 (1996) 73.
[88]   (a) Griem, R. H. (ed.). *Plasma Spectroscopy*. McGraw–Hill, New York, 1964; (b) Lochte-Holt-greven, W. (ed.). *Plasma Diagnostics*. North-Holland, Amsterdam, 1968.
[89]   Bernstein, L. S., and Zakin, M. R. *J. Phys. Chem.*, 99 (1995) 14619.
[90]   Frommhold, L. *SPIE Int. Soc. Opt. Eng.*, 3090 (1997) 272; (b) Frommhold, L., and Meyer, W. *AIP Conf. Proc.*, 9 (1997) 471.
[91]   Frommhold, L., and Atchley, A. A. *Phys. Rev. Lett.*, 73 (1994) 2883.
[92]   Hickling, R. *Phys. Rev. Lett.*, 73 (1994) 2853.
[93]   (a) Walton, A. J. *Adv. Phys.*, 26 (1977) 887; (b) Sweeting, L., and Rheingold, A. L. *J. Phys. Chem.*, 92 (1988) 5648; (c) Barsanti, M., and Maccarrone, F. *Riv. Nuovo Cimento*, 14 (1991) 1.
[94]   Schwinger, J. *Proc. Natl. Acad. Sci. USA*, 90 (1993) 2105.
[95]   (a) Eberlein, C. *Phys. Rev. A.*, 53 (1996) 2772; (b) Ibid. *Phys. Rev. Lett.*, 76 (1996) 3843.
[96]   (a) Lambrecht, A., Jaekel, M. -T., and Reynaud, S. *Phys. Rev. Lett.*, 78 (1997) 2267; (b) Esquivel-Sirvent, R., Jauregui, R., and Villareal, C. *Phys. Rev. A.*, 56 (1997) 2463.
[97]   Garcia, N., and Levanyuk, A. P. *JETP Lett.*, 64 (1996) 909.
[98]   (a) Prevelenslik, T. V. *Nucl. Sci. Tech.*, 7 (1996) 157; (b) Ibid. *Nucl. Sci. Tech.*, 8 (1997) 94.
[99]   Tsiklauri, D. *Phys. Rev. E.*, 56 (1997) 6245.
[100]   Mohanty, P., and Khare, S. V. *Phys. Rev. Lett.*, 80 (1998) 189.
[101]   Jauregui, R., Esquivel-Sirvent, R., and Villareal, C. *Phys. Rev. A.*, 57 (1998) 644.
[102]   Xu, N., Wang, L., and Hu, X. *Phys. Rev. E.*, 57 (1998) 1615.
[103]   Brenner, M. P., Lohse, D., Oxtoby, D., and Dupont, T. F. *Phys. Rev. Lett.*, 76 (1996) 1158.
[104]   Vuong, V. Q., and Szeri, A. J. *Phys. Fluids*, 8 (1996) 2354.
[105]   (a) Putterman, S. *Sci. Am.*, February (1995) 32; (b) Ibid. *Pour la Science*, 210 (1995) 42.
[106]   Barber, B. P., Hiller, R., Arisaka, K., Fetterman, H., and Putterman, S. *J. Acoust. Soc. Am.*, 91 (1992) 3061.
[107]   Lepoint, T., Lepoint-Mullie, F., and Henglein, A. in Crum, L. A. et al. (eds), Sonochemistry and Sonoluminescence. *NATO Series*, Kluwer, Dordrecht, 1999, pp. 285–290.
[108]   Barber, B. P., Weninger, K., Löfstedt, R., and Putterman, S. *Phys. Rev. Lett.*, 74 (1995) 5276.
[109]   Barber, B. P., and Putterman, S. *Phys. Rev. Lett.*, 69 (1992) 3839.
[110]   Van de Hulst, H. C. (ed.). *Light Scattering of Small Particles*. Wiley, New York, 1957.
[111]   Lentz, W. J., Atchley, A. A., Gaitan, D. F., and Maruyama, X. K. In Hobaeck, H. (ed.), *Advances in Nonlinear Acoustics*. World Scientific, London, 1993, pp. 400–405.
[112]   Lentz, W. J., Atchley, A. A., and Gaitan, D. F. *Appl. Opt.*, 34 (1995) 2648.
[113]   Tien, Y., Ketterling, J. A., and Apfel, R. E. *J. Acoust. Soc. Am.*, 100 (1996) 3976.
[114]   Weninger, K., Putterman, S. J., and Barber, B. P. *Phys. Rev. E.*, 54 (1996) 2205.
[115]   Prosperetti, A. in Ref. 30e, pp. 13–22.
[116]   Weissler, A., Cooper, H. W., and Snyder, S. J. *J. Am. Chem. Soc.*, 72 (1950) 1769.
[117]   Guyon, E., Hulin, J.-P., and Petit, L. (eds). *Hydrodynamique physique*. Intereditions, Paris, 1991.
[118]   Holt, R. G., Gaitan, D. F., Atchley, A. A., and Holzfuss, J. *Phys. Rev. Lett.*, 72 (1994) 1376.
[119]   Moran, M. J., Haigh, R. E., Lowry, M. E., Sweider, D. R., Abel, G. R., Carlson, J. T., Lewia, S. D., Atchley, A. A., Gaitan, D. F., and Maruyama, X. K. *Nuc. Instrum. Methods Phys. Res. B*, 96 (1995) 651.
[120]   Gompf, B., Günther, R., Nick, G., Pecha, R., and Eisenmenger W. *Phys. Rev. Lett.*, 79 (1997) 1405.
[121]   Barber, B. P., and Putterman, S. J. *Nature*, 352 (1991) 318.
[122]   Weninger, K. R., Barber, B. P., and Putterman, S. J. *Phys. Rev. Lett.*, 78 (1997) 1799.
[123]   Guenther, P., Heim, E., Schmitt, A., and Zeil, W. *Z. Naturforsch.*, 12A (1957) 521.
[124]   Guenther, P., Heim, E., and Eichkorn, G. *Z. Angew. Phys.*, 11 (1959) 274.
[125]   Wagner, W. U. *Z. Angew. Phys.*, 10 (1958) 445.
[126]   Jarman, P. D. *Proc. Phys. Soc. B*, 72 (1959) 628.

[127] Golubnichii, P. I., Gondrakov, V. D., and Protopopov, K. V. *Sov. Phys. Acoust.*, 15 (1970) 464; Ibid, 16 (1970) 115.

[128] Taylor, K. J., and Jarman, P. D. *Aust. J. Phys.*, 23 (1970) 319.

[129] (a) Negishi, K. *J. Phys. Soc. Japan*, 16 (1961) 1450; (b) Ibid. *Acustica*, 10 (1960) 124.

[130] McLeay, R. Q., and Holroyd, L. F. *J. Appl. Phys.*, 32 (1961) 449.

[131] Matula, T. J., Roy, R. A., and Mourad, P. *J. Acoust. Soc. Am.*, 101 (1997) 1994.

[132] Barber, B. P., Löfstedt, R., and Putterman, S. *J. Acoust. Soc. Am.*, 89 (1991) S1885.

[133] Barber, B. P., Wu, C. C., Löfstedt, R., Roberts, P. H., and Putterman, S. *J. Phys. Rev. Lett.*, 72 (1994) 1380.

[134] Bertholet, T., Lepoint, T., and Lepoint-Mullie, F. In preparation.

[135] Hiller, R., Putterman, S. J., and Barber, B. P. *Phys. Rev. Lett.*, 69 (1992) 1182.

[136] Löfstedt, R., Barber, B. P., and Putterman, S. *J. Phys. Fluids A*, 5 (1993) 2911.

[137] Löfstedt, R., Weninger, K., Putterman, S. J., and Barber, B. P. *Phys. Rev. E*, 51 (1995) 4400.

[138] Holt, R. G., and Gaitan, D. F. *Phys. Rev. Lett.*, 77 (1996) 3791.

[139] (a) Brenner, M. P., Lohse, D., and Dupont, T. F. *Phys. Rev. Lett.*, 75 (1995) 954; (b) Hilgenfeldt, S., Lohse, D., and Brenner, M. P. *Phys. Fluids*, 8 (1996) 2808; (c) Kondic, L., Yuan, C., and Chan, C. K. *Phys. Rev. E*, 57 (1998) 32.

[140] Hiller, R. A., Weninger, K., Putterman, S. J., and Barber, B. P. *Phys. Rev. Lett.*, 75 (1995) 3549.

[141] (a) Lohse, D., Brenner, M. P., Dupont, T. F., Hilgenfeldt, S., and Johnston, B. *Phys. Rev. Lett.*, 78 (1997) 1359; (b) Lohse, D., and Hilgenfeldt, S. *J. Chem. Phys.*, 107 (1997) 6986.

[142] Matula, T. J., and Crum, L. A. *Phys. Rev. Lett.*, 80 (1998) 865.

[143] Hiller, R. A., and Putterman, S. J. *Phys. Rev. Lett.*, 75 (1995) 3549.

[144] Hiller, R. A., and Putterman, S. J. *Phys. Rev. Lett.* 77 (1996) 2345.

[145] Flint, E. B., and Suslick, K. S. *J. Phys. Chem.*, 95 (1991) 1484.

[146] Hiller, R., Weninger, K., Putterman S. J., and Barber, B. P. *Science*, 266 (1994) 248.

[147] Weninger, K., Hiller, R., Barber, B. P., Lacoste, D., and Putterman S. J. *J. Phys. Chem.*, 99 (1995) 14195.

[148] Matula, T. J., Roy, R. A., Mourad, P., McNamara, W. B., and Suslick, K. S. *J. Acoust. Soc. Am.*, 75 (1995) 2602.

[149] Seghal, C., Sutherland, R. G., and Verrall, R. E. *J. Phys. Chem.*, 84 (1980) 525.

[150] Didenko, Y. T., Natisch, D. N., Pugach, Y. A., Polovinka, Y. A., and Kvochka, V. I. *Ultrasonics*, 32 (1994) 71.

[151] Hiller, R. A., and Putterman, S. J. *J. Acoust. Soc. Am.*, 100 (4, Pt. 2) (1996) 2717.

[152] Pearse, R. W. B., and Gaydon, A. G. (eds.). *The Identification of Molecular Spectra*. Chapman & Hall, London, 1963, pp. 94–97.

[153] Flint, E. B., and Suslick, K. S. *J. Am. Chem. Soc.*, 111 (1989) 6987.

[154] Fortov, V. E. *Sov. Phys. Usp.*, 25 (1982) 781.

[155] Weninger, K., Cho, H., Hiller, R. A., Putterman, S. J., and Williams, G. A. *Phys. Rev. E*, 56 (1997) 6745.

[156] (a) Didenko Y. T., and Pugach S. P. *Adv. Non Linear Acoust.*, (1993) 412; (b) Ibid. *Ultrasonics Sonochem.*, 1 (1994) S9; (c) Ibid. *J. Phys. Chem.*, 98 (1994) 9743.

[157] Sehgal, C., Sutherland, R. G., and Verrall, R. E. *J. Phys. Chem.*, 84 (1980) 388.

[158] Didenko, Y. T., Pugach, S. P., and Gordeichuk, T. V. *Opt. Spectrosc.*, 80 (1996) 821.

[159] Günther, P., Zeil, W., Grisar, U., and Heim, E. Z. *Elektrochem.*, 61 (1957) 188.

[160] Didenko, Y. T. *Acoust. Phys.*, 43 (1997) 215.

[161] Vallas-Dubois, S., Haug, R., and Prudhomme, R. O. *J. Chim. Phys.*, 73 (1978) 855.

[162] Ayad, M. *Infrared Phys.*, 11 (1971) 249.

[163] Sehgal, C., Sutherland, R. G., and Verrall, R. E. *J. Phys. Chem.*, 84 (1980) 396.

[164] Becker, L., Bada, J. L., Kemper, K., and Suslick, K. S. *Mar. Chem.*, 40 (1992) 143.

[165] Matula, T. J., Roy, R. A., and Mourad, P. *Phys. Rev. Lett.*, 75 (1995) 2602.

[166] Sehgal, C., Steer, R. P., Sutherland, R. G., and Verrall, R. E. *J. Chem. Phys.*, 70 (1979) 2242.

[167] Heim, E. Z. *Angew. Phys.*, 12 (1960) 423.
[168] Sehgal, C., Sutherland, R. G., and Verrall, R. E. *J. Phys. Chem.*, 84 (1980) 529.
[169] Sehgal, C., Sutherland, R. G., and Verrall, R. E. *J. Phys. Chem.*, 84 (1980) 227.
[170] Sehgal, C., Sutherland, R. G., and Verrall, R. E. *J. Phys. Chem.*, 85 (1981) 315.
[171] Suslick, K. S., and Flint, E. B. *Nature*, 330 (1987) 553.
[172] Suslick, K. S., Flint, E. B., Grinstaff, M. W., and Kemper, K. A. *J. Phys. Chem.*, 97 (1993) 3098.
[173] Flint, E. B., and Suslick, K. S. *Science*, 253 (1991) 1397.
[174] Suslick, K. S., and Kemper, K. A. In Blake, J. R. et al. (eds.), *Bubble Dynamics and Interface Phenomena*. Kluwer Academic, Dordrecht, 1994, pp. 311–320.
[175] Suslick, K. S., and Kemper, K. A. *Ultrasonics*, 31 (1993) 463.
[176] (a) Carlson, J. T., Lewia, S. D., Atchley, A. A., Gaitan, D. F., Lowry, M. E., Moran, M. J., and Sweider, D. R. Non Linear Acoustics, 13th ISNA, Bergen Norway, ed. H. Hoback, London, 1993 p. 406; (b) Gaitan, D. F., Atchley, A. A., Lewia, S. D., Carlson, J. T., Maruyama, X. K., Moran, M., and Sweider, D. *Phys. Rev. E*, 54 (1996) 525.
[177] Crum, L. A. *Phys. Today*, Sept. (1994) 22.
[178] Henglein, A. in Ref. 20, Vol. 3, pp. 17–83.
[179] (a) Lepoint-Mullie, F., De Pauw, D., Lepoint, T., Supiot, P., and Avni, R. *J. Phys. Chem.*, 100 (1996) 12138; (b) a more accurate analysis taking into consideration the reabsorption of the plasma is in preparation; An erratum accompanies Ref. 179a in *J. Phys. Chem.*, Issue 04/29/1999.
[180] Casimir, H. B. G., and Polder, D. *Phys. Rev.*, 73 (1948) 360.
[181] Unnikrishnan, C. S., and Mukhopadhyay, S. *Phys. Rev. Lett.*, 77 (1996) 4690.
[182] Hickling, R., and Plesset, M. S. *Phys. Fluids*, 7 (1964) 7.
[183] Matula, T. J., Hallaj, I. M., Cleveland, R. O., Crum, L. A., Moss, W. C., and Roy, R. A. *J. Acoust. Soc. Am.*, 103 (1998) 1377.
[184] Ohl, C. D., Lindau, O., and Lauterborn, W. *Phys. Rev. Lett.*, 80 (1998) 393.
[185] Chen, S.-Y., and Takeo, M. *Rev. Mod. Phys.*, 29 (1957) 20.
[186] Lewshin, V. L., and Rschevkin, S. N. *C. R. Acad. Sci.*, 16 (1937) 399.
[187] Balzani, V., and Scandola, F. (eds.). *Supramolecular Photochemistry*. Ellis Horwood, Chichester, 1991, pp. 237–238.

OH RADICAL FORMATION AND DOSIMETRY IN THE SONOLYSIS OF AQUEOUS SOLUTIONS

Clemens von Sonntag, Gertraud Mark,
Armin Tauber, and Heinz-Peter Schuchmann

Advances in Sonochemistry
Volume 5, pages 109–145.
Copyright © 1999 by JAI Press Inc.
All rights of reproduction in any form reserved.
ISBN: 0-7623-0331-X

1. INTRODUCTION

The action of ultrasound on aqueous solutions, as on other liquids, subjects each volume element of the liquid to a succession of phases of compressive and tensile stress. Depending on the frequency and intensity of the acoustic wave-field, as well as on the content of dissolved gas, tensile stress can lead to the rupture of the liquid and the formation of a bubble in the stressed volume element ("cavitation"). The bubble is initially under reduced pressure and fills with gas and other volatile solutes that emanate from the liquid, together with water vapor. The partitioning of these volatiles between the liquid and the gas phases is expected to be very efficient since a depletion of these substances in the boundary layer of the bubble is prevented by efficient mixing due to sound-induced convection that operates around the bubble. The external pressure, assisted by the subsequent phase of compressive stress and the effect of surface tension, forces the bubble to shrink or to collapse; this process is more or less adiabatic, so that very high temperatures (a few thousand degrees) can be reached in the compression phase. These are at the origin of the formation of various free radicals from the gas and vapor mixture in the bubble. Some of these free radicals escape from the bubble into the solution.

Among them, the OH radical is of particular interest because it is one of the most reactive free-radical species known, and any application of ultrasound to aqueous media must take into account its possible reactions with the material that is dissolved in such media. It is therefore useful to know the yield of OH radicals that become manifest in the solution, for which reliable OH radical-specific dosimeters are necessary.

OH radicals are formed on thermolysis of H_2O [1–3] in the collapsing bubble. Besides the homolytic dissociation into H and OH free radicals [reaction (1)] one may envisage a decomposition into H_2 and an O atom [reaction (2)] [4–8]. Atomization ($H_2O \rightarrow 2$ H· + O) is not expected to play a part since the energy requirement at 930 kJ mol^{-1} is almost twice as high as that for reaction (1) or (2). In water vapor at about 2000 K, the proportion of water molecules decomposed approaches the parts per hundred range [9,10].

$$H_2O \rightarrow H· + ·OH \qquad (1)$$

$$H_2O \rightarrow H_2 + O \qquad (2)$$

The O atoms are largely converted into OH radicals [reaction (3)]. At the high temperature in the interior of the bubble, endothermic reactions such as reaction (4) ($\Delta H = 63$ kJ mol^{-1}) [11] can readily occur and, in the absence of volatiles other than H_2O that could react with H· and/or ·OH, an excess of ·OH over H· will be the result.

One estimate has been that about five times more OH radicals than H atoms reach the boundary layer (initial concentrations, 5×10^{-3} versus 1×10^{-3} mol dm^{-3}) [3].

$$O + H_2O \rightarrow \cdot OH + \cdot OH \tag{3}$$

$$H^{\cdot} + H_2O \rightarrow H_2 + \cdot OH \tag{4}$$

In the absence of volatile OH-radical scavengers, the product H_2 accumulates only to a small degree as it is consumed by a process reverse to reaction (4). To a large extent, these radicals react with one another in the gas phase and beyond [reactions (5)–(7)]; of those that reach the boundary layer of the bubble some will pass on into the solution.

$$\cdot OH + \cdot OH \rightarrow H_2O_2 \tag{5}$$

$$\cdot OH + H\cdot \rightarrow H_2O \tag{6}$$

$$H\cdot + H\cdot \rightarrow H_2 \tag{7}$$

Owing to the relatively low strength of the $-O-O-$bond (213 kJ mol^{-1}), the thermal stability range of H_2O_2 is restricted. In fact, a large part of the H_2O_2 that is observed afterward is expected to be produced by reaction (5) not in the gas phase but in the liquid boundary layer that surrounds the bubble.

The product H_2O_2 is of particular interest since it is a compound that constitutes, as it were, the scent that the OH radicals leave on their trail from the collapsing bubble into the liquid, to the extent that H_2O_2 is formed by reaction (5). On the other hand, reactions (8)–(12) in total diminish the H_2O_2 output.

$$\cdot OH + H_2O_2 \rightarrow H_2O + HO_2^{\cdot} \tag{8}$$

$$\cdot H + H_2O_2 \rightarrow H_2O + \cdot OH \tag{9}$$

$$\cdot OH + HO_2^{\cdot} \rightarrow H_2O + O_2 \tag{10}$$

$$H\cdot + O_2 \rightarrow HO_2^{\cdot} \tag{11}$$

$$HO_2^{\cdot} + HO_2^{\cdot} \rightarrow H_2O_2 + O_2 \tag{12}$$

It is seen that these reactions lead to the production of oxygen, which may introduce peroxyl-radical reaction pathways and thereby complicate the sonolysis mechanism of certain solutes.

As the OH radicals pass into the liquid phase, their initial distribution is nonhomogeneous as they are cumulated in the boundary layer of the bubble. Recombination may therefore be favored over the reaction with a solute. This phenomenon of spatial nonhomogeneity is well known in radiation chemistry (see [12–17]), where the energy of the ionizing radiation is *also* deposited discontinuously; an individual ionization and excitation event leads to the production of one or more pairs of radicals in close proximity. These radicals undergo "geminate" recombination, in competition with their diffusion apart, into the bulk of the solution. In radiolysis, suitable scavengers at sufficiently high concentration allow us to probe the regions of geminate radical pairs (these regions are called *spurs*). The same approach can in principle be used to distinguish between the H_2O_2 formed in reaction (5) in the domain that is accessible to the scavenger, i.e., the bulk of the liquid phase and the boundary liquid/bubble, and the rest, if any. However, experience from radiolysis shows that high scavenger concentrations (up to about molar) must be used to approach complete suppression of geminate recombination. Reliable computational models of the nonhomogeneous kinetics are available for the case of ionizing radiation [14,18], but a model that describes the chemical phenomena observed with ultrasound to a similar degree of adequacy is lacking to date. In contrast to radiation chemistry, sonochemistry encompasses chemical reactions in a gaseous as well as a liquid phase, and the question might justifiably be posed as to whether a model that is accurate *in detail* will ever be feasible, given that the dynamics of cavitation-bubble behavior depends in a complicated manner on a considerable number of parameters, especially the frequency and intensity of the sonic field, on the vapor pressure of the liquid, and on the type and concentration of the dissolved gas. Nevertheless, a kinetic model that emphasizes the gas-phase pyrolysis aspects of the sonolytic decomposition of aromatic hydrocarbons in aqueous solution has been attempted recently [19]. The complexities of the cavitation phenomenon have been discussed (see [8,20–34]).

One of the consequences of the interplay of the experimental parameters is that two distinct modes of bubble evolution, called *stable* and *transient* cavitation, are encountered. In the former, the size of the bubble oscillates a number of times, repeatedly going through a phase of moderate maximal dilation, followed by compression with a corresponding rise in temperature. Indications are that in the case of stable cavitation the compression of the bubble is often far from adiabatic (see [29, p. 67]). In transient cavitation, a more pronounced dilation of the bubble means that the inertial force that develops during contraction is greater; this leads to a final collapse where the degrees of maximal compression and heating are much higher, with the adiabatic situation more closely approximated. Hence, the chemical effects with regard to the gaseous contents of the bubble must differ considerably between the two.

In order to assess the sonolytic OH radical yield, it is useful to follow the practical approach established in radiation chemistry. Most of these experiments in aqueous radiation chemistry are carried out at low scavenger concentrations (ca. 10^{-3} mol dm^{-3}). In radiation chemistry, the yields of H_2O_2 and H_2 (these are formed by geminate recombination from the respective radicals) do not vary significantly on varying the scavenging capacity [15] (i.e., the product of the scavenger concentration and the rate constant; dimension: s^{-1}) within a wide range. Conversely, this applies to the yields of H atoms and OH radicals that become manifest in the bulk of the solution [35].

In radiation chemistry, well-established dosimeters are available, e.g., the Fricke dosimeter for γ-radiolysis [36], and thiocyanate dosimetry for pulse radiolysis [37,38]. Since the chemistries of these dosimeters are well understood, the yield of OH radicals can be derived from such measurements. While for the most part homogeneous kinetics can be applied in radiation chemistry of sparsely ionizing radiation, i.e., γ-rays and fast electrons, since here the clusters of geminate radicals are relatively small and tenuous, the situation in the sonolytic systems is more complex: The nonhomogeneity of sonolytic systems regarding free-radical production is much more pronounced since the number of free radicals "geminately" produced per bubble-collapse event far exceeds that produced by this type of radiation in a single ionization event. Evidently, the spatial scale of the cavitation phenomenon is microscopic not molecular, in contrast to that of an ionization event, which is several orders of magnitude smaller. An additional complexity is introduced by the transient existence of a hot gas-phase environment where a solute that has emanated from the solution into the bubble, or which may have been vaporized in the course of a thermal ablation affecting the boundary layer, suffers thermolytic decomposition whose products may then obscure the free-radical chemistry going on in the liquid.

While the relative yields of products formed from volatile solutes may vary according to the power and the frequency of the insonation, this is not expected to be the case with those solutes that are strictly confined to the liquid phase. Their sonolysis occurs in the solution and is mainly induced by the OH radical, regardless of the conditions of cavitation. In the absence of volatile radical scavengers, the OH radical (total reaching the liquid) predominates [3,39] over the H atom in a proportion of about 4:1 or 5:1 [40] because of reaction (4). One might expect that in the course of the heating of the bubble's contents, the boundary layer is vaporized to a depth of several molecular diameters, or that in the event of nonspherical transient cavitational collapse, tiny droplets of liquid may become detached from the boundary of the bubble and vaporized (see [29]); thus, even nonvolatile solutes could enter the gas phase and suffer some degree of thermolysis. It is thought that this effect might in fact determine the sonochemistry observed in the case of very concentrated solutions [3]. The observation of alkali-atom optical emission in the luminescence spectra of salt solutions (see [2; 31, p. 237; 41]) a very sensitive indicator, points in this direction. It will be shown below that at low solute concentrations this effect is so small as to be unobservable as far as the formation of products is concerned.

In the main, three OH radical dosimeters are currently employed in aqueous sonochemistry. For their adequate use it is important to understand the underlying chemistry, which is discussed in the present review.

2. DOSIMETERS IN CURRENT USE

The OH radical yield per unit sonic energy input varies according to the equipment used. For this reason, beyond the measurement [42] of the sonic energy absorbed in the insonated medium, it is necessary to separately determine the extent of chemical change by means of a chemical actinometer. For the quantification of OH radicals that have reached the liquid, solutes should be used that do not evaporate, i.e., that cannot intercept the OH radicals in the gas phase, or might interfere with OH radical formation by partly forestalling the temperature rise in the shrinking gas bubble on account of a high heat capacity, and endothermic decomposition. Only fully dissociated electrolytes meet this condition.

Dosimeters that fulfill this requirement are the Fricke dosimeter (oxidation of Fe^{2+} in acid solution) [3,43–46], the terephthalate dosimeter (formation of 2-hydroxyterephthalate) [45,47,48], and the iodide dosimeter (formation of io-dine/I_3^-) [49–52]. However, the Fricke dosimeter, besides reacting with OH radicals, also responds to hydrogen peroxide, H atoms, and HO_2 radicals. The terephthalate dosimeter detects only the OH radicals, but an additional oxidant is required; the yield of the indicator product 2-hydroxyterephthalate varies strongly with the nature of the oxidant (e.g., O_2 versus $IrCl_6^{2-}$). It is typically used at a terephthalate concentration of 2×10^{-3} mol dm^{-3}, i.e., at a low scavenging capacity. The iodide dosimeter has been used also at relatively high iodide concentrations.

Iodide dosimetry is sometimes done in the presence of carbon tetrachloride (Weissler reaction) [53]; the I_2 that is measured under these conditions does not only originate from the action of OH radicals but also from chlorine atoms and Cl_2 that are formed in the gas-phase decomposition of carbon tetrachloride [54–56]. Evidently, this system does not conform to the restriction to purely liquid-phase-reaction actinometry.

2.1 Fricke Dosimeter

In the Fricke dosimeter (1×10^{-3} mol dm^{-3} $FeSO_4$ in 0.4 mol dm^{-3} H_2SO_4, oxygen- or air-saturated; higher ferrous sulfate concentrations have also been used [40]), Fe^{2+} is oxidized by OH radicals and hydrogen peroxide according to reactions (13) and (14) and the Fe^{3+} determined. The H atoms are scavenged by O_2 [reaction (15)]. At this low pH, Fe^{2+} is also oxidized by the HO_2 radical, whereby hydrogen peroxide is formed [reaction (16)]. This in turn reacts with Fe^{2+} to give rise to an OH radical.

$$\cdot OH + Fe^{2+} + H^+ \rightarrow H_2O + Fe^{3+} \tag{13}$$

$$H_2O_2 + Fe^{2+} + H^+ \rightarrow H_2O + Fe^{3+} + \cdot OH \tag{14}$$

$$H\cdot + O_2 \rightarrow HO_2 \tag{15}$$

$$HO_2^- + Fe^{2+} + H^+ \rightarrow H_2O_2 + Fe^{3+} \tag{16}$$

Thus, two Fe^{3+} are generated per molecule of hydrogen peroxide and accordingly three Fe^{3+} by one HO_2 radical [Eq. (17); the symbol G denotes the energy-specific yield, "G value," dimension: mol J^{-1}; the energy of the ultrasound absorbed is determined by calorimetry with the insonated vessel sufficiently insulated [40,42,51]].

$$G(Fe^{3+})_{O_2} = G(\cdot OH) + 2\, G(H_2O_2) + 3\, G(HO_2^-) \tag{17}$$

From Eq. (17) it is obvious that the formation of Fe^{3+} is not a direct measure of the free OH radical yield, and that additional experimentation is required to determine the contributions of the H_2O_2 and the H atom (HO_2 radical). The contribution of the latter can be evaluated by conducting the Fricke dosimetry in the presence of Cu^{2+} ions ("copper-Fricke") [46,57]. Cu^{2+} (1×10^{-2} mol dm^{-3}) readily scavenges HO_2 radicals according to reaction (18), while Cu^+ thus formed regenerates Fe^{2+} by reducing Fe^{3+} [reaction (19)]:

$$HO_2^- + Cu^{2+} \rightarrow O_2 + Cu^+ + H^+ \tag{18}$$

$$Cu^+ + Fe^{3+} \rightarrow Cu^{2+} + Fe^{2+} \tag{19}$$

As a consequence, the yield of Fe^{3+} is now given by

$$G(Fe^{3+})_{Cu,\,O_2} = G(\cdot OH) + 2\, G(H_2O_2) - G(HO_2^-) \tag{20}$$

The difference between Eqs. (17) and (20) is a measure of $G(HO_2^-)$, i.e., $G(H\cdot)$.

It is thus apparent that Fricke dosimetry allows the separate determination of $G(\cdot H/HO_2^-)$ and the sum of $G(\cdot OH) + 2G(H_2O_2)$. The latter term is practically independent of the scavenging capacity, except when this is very small (see Figures 2–4 below) because under these conditions some of the H_2O_2 is destroyed in reaction (8). As the scavenging capacity increases, OH radicals increasingly react with the scavenger and the amount of H_2O_2, essentially produced by $\cdot$OH recombination [reaction (5)], declines in proportion.

For a given concentration of substrate, the dosimeter system should be used such that the scavenging capacities are the same in both solutions. This is more important in sonolysis than in radiolysis. Over the range of low scavenging capacities, in sonolysis $G(\cdot OH)$ rises more steeply as a function of scavenging capacity than in radiolysis, the reason being that a cluster of "geminate" OH radicals (which tend to form H_2O_2 by recombination if not intercepted) in sonolysis is larger and comprises far more OH radicals than in electron-beam or γ-radiolysis.

Ferrous sulfate dosimetry can also be carried out in oxygen-free, i.e., argon saturated, solution. The H atom *oxidizes* the Fe^{2+} [reaction (21), $k = 7.7 \times 10^7$ mol $dm^{-3}\,s^{-1}$], while it *reduces* the Cu^{2+} [reaction (22), $k = 9.7 \times 10^7$ mol $dm^{-3}\,s^{-1}$]:

$$Fe^{2+} + H\cdot + H^+ \rightarrow Fe^{3+} + H_2 \tag{21}$$

$$Cu^{2+} + H\cdot \rightarrow Cu^+ + H^+ \tag{22}$$

In this case the Fe^{3+} yields are given by (see [40])

$$G(Fe^{3+})_{Ar} = G(\cdot OH) + 2\,G(H_2O_2) + G(H\cdot) \tag{23}$$

$$G(Fe^{3+})_{Cu,\,Ar} = G(\cdot OH) + 2\,G(H_2O_2) - G(H\cdot) = G(Fe^{3+})_{Cu,\,O_2} \tag{24}$$

2.2 Terephthalate

The terephthalate dosimeter makes use of the OH-radical-induced hydroxylation of the terephthalate ion. Like the Fricke dosimeter, it was originally developed for the dosimetry of ionizing radiation [58,59]. The indicator product is 2-hydroxyterephthalate, which is readily determined on the basis of its fluorescence. The mechanism that leads to this product is complex and has been elucidated in depth recently [60]. It is readily adaptable to the sonolytic situation. The key aspects of the mechanistic study are repeated here. The OH radicals add at an almost diffusion-controlled rate ($k = 3.3 \times 10^9$ dm^3 mol^{-1} s^{-1}), preferentially (85%) in the 2-position [reaction (25)]; the remainder (15%) enters *ipso* to the position of the carboxylate function. In the absence of an oxidant, the yield of 2-hydroxyterephthalate is very low because the disproportionation of hydroxycyclohexadienyl-type radicals is ineffective compared with their recombination. In contrast to other hydroxycyclohexadienyl-type radicals that can be readily oxidized by $Fe(CN)_6^{3-}$ [61], the one derived from terephthalic acid requires a more potent oxidant. $IrCl_6^{2-}$ has been found to oxidize these radicals quantitatively and at a fast rate [reaction (26); $k = 7.7 \times 10^7$ dm^3 mol^{-1} s^{-1}]. O_2 as an oxidant does not react quantitatively. The reason for this is that HO_2^- elimination [reaction (29); $k = 390$ s^{-1}] subsequent to O_2 addition is incomplete, rather than the slowness of this addition [reactions (27) and (28); $k = 1.6 \times 10^7$ dm^3 mol^{-1} s^{-1}] or its reversibility [reactions (−27) and (−28); $k = 3.4 \times 10^3$ s^{-1}] (for details see [62]). Ring closure of the peroxyl radicals, further O_2 addition, and bimolecular decay of these peroxyl radicals leads to a large number of fragmentation products [reactions (30/−30) and (31)]. Their nature has not been elucidated in the case of terephthalic acid, but other aromatic systems, e.g., benzene [63] and chlorobenzene [64] that behave in a similar manner, have been studied in detail. Thus, with O_2, the yield of 2-hydroxyterephthalate is only 35% of the OH radical yield, while in the presence of $IrCl_6^{2-}$ it is 85%. The fact that the terephthalate dosimeter works without O_2 enlarges its scope since, employing $IrCl_6^{2-}$ as oxidant, it can be applied in the sonolysis of solutions saturated with a noble gas or nitrogen.

Owing to the fact that the generation of the OH radicals in the sonicated aqueous solution is nonhomogeneous, the yield of OH free radicals observed will be higher the higher the terephthalate concentration. This exposes a drawback of the terephthalate system. $IrCl_6^{2-}$ is somewhat unstable toward terephthalate. Even though the products of this reaction do not detract from the actinometry proper, terephthalate if used in too high a concentration (limit: between 10^{-2} and 5×10^{-2} mol dm^{-3}) might cause the oxidant to be consumed before the sonolysis experiment is over. In the concentration range below 10^{-2} mol dm^{-3}, however, the terephthalate actinometer is reliable. Figure 1 shows how the yield of 2-hydroxyterephthalate

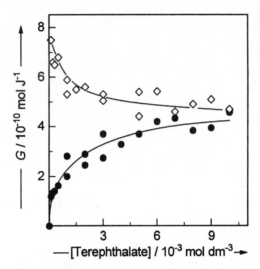

Figure 1. Sonolysis of Ar-saturated terephthalate solutions in the presence of 2×10^{-4} mol dm^{-3} IrCl$_6^{2-}$ (321 kHz, 170 W kg^{-1}). Yields of H$_2$O$_2$ ($\Diamond$) and OH radicals ($\bullet$, measured value times 0.5) which are free to react with solute, as a function of the terephthalate concentration.

increases with increasing terephthalate concentration, while that of H$_2$O$_2$ decreases in equal measure [40], as the influence of the OH radical scavenger reaches more deeply toward the more concentrated regions of the OH radical cluster and increasingly prevents recombination [reaction (5)]. This illustrates the point made above that most of the H$_2$O$_2$ produced arises from the recombination of the geminately produced OH radicals.

Another complexity of the terephthalate dosimeter must be mentioned. The iodometric assay [65] of the sonolyzed solution for hydrogen peroxide reveals the presence of another peroxidic product in low yield [40]. It is known that hydrogen peroxide may react much faster than organic hydroperoxides or peroxides; thus, the assay permits a differentiation between the former and the rest by monitoring the buildup of I$_3^-$. In the present case the slow component is probably due to hydroperoxidic and/or endoperoxidic material [40].

2.3 Iodide

The OH radical oxidizes iodide ion at a practically diffusion-controlled rate [reaction (32); $k = 1.1 \times 10^{10}$ dm^3 mol^{-1} s^{-1} [66]], giving rise to an iodine atom. This forms a complex with an iodide ion [equilibrium (33)]; the I$_2^-$ radicals decay bimolecularly yielding iodine [reaction (34)], which forms the I$_3^-$ complex with iodide [reaction (35)], stability constant $K = 725$ dm^3 mol^{-1}. I$_3^-$ is the species that is monitored (see [40]).

$$\cdot OH + I^- \rightarrow OH^- + I\cdot \tag{32}$$

$$I\cdot + I^- \rightleftharpoons I_2^- \tag{33}$$

$$2\,I_2^- \rightarrow I_2 + 2\,I^- \tag{34}$$

$$I_2 + I^- \rightleftharpoons I_3^- \tag{35}$$

Some of the H atoms formed in relatively low yields alongside the OH radical may react with the I_2 and thus reduce its yield [reaction (36); $k = 3.5 \times 10^{10}$ dm^3 mol^{-1} s^{-1} [66]], as may the superoxide radical [reaction (37); $k = 6 \times 10^9$ dm^3 mol^{-1} s^{-1} [67]].

$$H\cdot + I_2 \rightarrow HI + I\cdot \tag{36}$$

$$O_2^{\cdot -} + I_2 \rightarrow O_2 + I_2^{\cdot -} \tag{37}$$

Hydrogen peroxide reacts very slowly with iodide in an uncatalyzed reaction [reaction (38)], but with molybdate as a catalyst [65] reaction (38) proceeds with a half-life of 0.33 s at an iodide concentration of 0.2 mol dm^{-3} (the rate depends nonlinearly on the iodide concentration) [68]. This feature allows the OH radical yield *and* the H_2O_2 yield to be determined from the same sample:

$$H_2O_2 + 2\,I^- + 2\,H^+ \rightarrow I_2 + 2H_2O \tag{38}$$

when the I_2 yield has been measured (as I_3^-, *vide supra*) immediately after sonolysis, catalyst is added; the increment in the I_2 yield is then attributed to the H_2O_2. Care must be taken since in neutral or basic solutions I_3^- destroys H_2O_2 by oxidation. For this reason the experiments should be done under slightly acidic conditions (e.g., pH 5.5), as otherwise, especially at low iodide concentrations and high H_2O_2 yields, the I_2 yield is found to be too low [49].

The dependence of the I_2 and H_2O_2 yields on iodide concentration have been measured following this approach. It has been found that at low iodide concentrations, the yield of I_2 at first rises strongly (Figure 2), to gradually flatten off (Figure 3), but very high iodide concentrations are required to fully suppress the formation of H_2O_2 (Figure 4) [49].

Other scavengers instead of iodide, e.g., terephthalate, influence the H_2O_2 yield in a like manner at similar scavenging capacities, as illustrated in Figure 5. This proves the point made above, namely, that in sonolysis one is dealing with nonhomogeneous reaction kinetics.

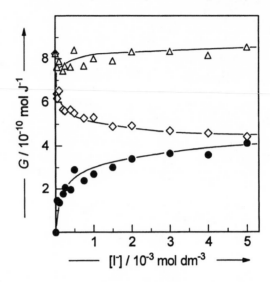

Figure 2. Sonolysis of Ar-saturated aqueous solutions of iodide (321 kHz; 170 W kg^{-1}). Yields of I_2 (●), H_2O_2 (◇), and $I_2 + H_2O_2$ (△) as a function of the iodide concentration at low iodide concentrations.

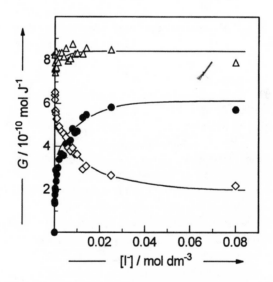

Figure 3. Sonolysis of Ar-saturated aqueous solutions of iodide (321 kHz; 170 W kg^{-1}). Yields of I_2 (●), H_2O_2 (◇), and $I_2 + H_2O_2$ (△) as a function of the iodide concentration at medium iodide concentrations.

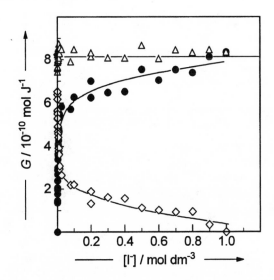

Figure 4. Sonolysis of Ar-saturated aqueous solutions of iodide (321 kHz; 170 W kg^{-1}). Yields of I$_2$ (●), H$_2$O$_2$ (◇), and I$_2$ + H$_2$O$_2$ (△) as a function of the iodide concentration at high iodide concentrations.

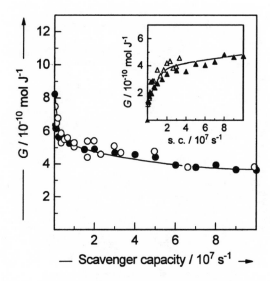

Figure 5. Dependence of the yields of H$_2$O$_2$ (○, ●) and OH radicals free to react with a solute (△, ▲; $G(\cdot OH) \times 0.5$) on the scavenging capacity [= $k(\cdot OH + S) \times [S]$; ○, △ terephthalate, $k = 3.3 \times 10^9$ dm^3 mol^{-1} s^{-1} [60] and ●, ▲ iodide, $k = 1.1 \times 10^{10}$ dm^3 mol^{-1} s^{-1} [49]].

3. FACTORS AFFECTING THE OH RADICAL YIELD

3.1 Power and Frequency

It is well known that the cavitation process shows a threshold with respect to sonic intensity (see [29]). The OH radical production then increases with increasing absorbed ultrasound power, finally to flatten out. Some results from our laboratory are shown in Figure 6.

It is further seen that there is a dependence on ultrasound frequency as well. While the level of the plateau goes through a maximum between about 300 kHz and 1 MHz, the threshold increases with the frequency throughout the range of frequencies tested, in agreement with expectation (see [31; 69, p. 206]). A similar efficiency maximum along the frequency scale has been observed for various products by us [40] and others (see [70]). The existence of this maximum has to do with the fact that the cavitation bubbles tend to be smaller at higher frequencies and therefore less energetic [31]. The maximum is also reminiscent of the efficiency maximum of ultrasound transduction [40] which may be a characteristic of cavitating dilute aqueous systems under these pressure and temperature conditions.

Since a homogeneous sonic field is not realized in the solutions under the usual experimental conditions [42,71,72], the dependence on frequency and power implies that the detailed aspect of the yield versus power and yield versus frequency curves must differ according to the insonation setup. The distribution of the more

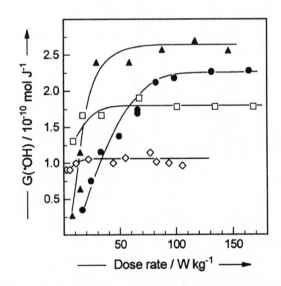

Figure 6. Sonolysis of air-saturated water. Free OH radical yield (determined with 2 × 10^{-3} mol dm^{-3} terephthalate) as a function of the dose rate at different sonication frequencies ($\Diamond$, 169 kHz; $\square$, 321 kHz; $\blacktriangle$, 585 kHz; $\bullet$, 1040 kHz).

and the less intensely insonated domains within the volume of the liquid even in the case where the transducer plate forms the bottom of a cylindrical vessel [40] will change with the power and the frequency owing to internal reflections at the walls of the vessel. Variations will also occur as a result of any change in the distribution of radiance across the surface of the oscillator. With such changes the relative importance of stable and transient cavitation will also vary. The curves in Figure 6 therefore represent volume-average values. It turns out, on a sampling of different regions of the sonication-bath volume (500 mL), that the characteristics of the free-OH-radical and H_2O_2 yield-versus-electrical-input curves differ remarkably (Figure 7). When the sonolysis is carried out in the sonication bath directly, the yield rises monotonically and appears to reach a plateau at relatively low values. In a small probing vessel (8 mL liquid volume) positioned axially in the cylindrical sonication bath, a steep maximum is observed at low power-input rates; with the probing vessel positioned off-center the behavior is qualitatively similar but with yields lower and the maximum displaced toward higher input power. This is shown in Figure 7 where the abscissa is scaled in terms of electric-input readings. This behavior is observed under air *and* under argon, i.e., it does not depend on the kind

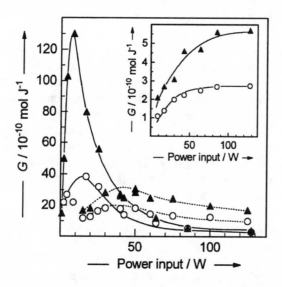

Figure 7. Energy yields of free OH radicals (o, determined in the presence of 2×10^{-3} mol dm^{-3} terephthalate and 2×10^{-4} mol dm^{-3} $IrCl_6^{2-}$) and H_2O_2 (▲, in the presence of 2×10^{-3} mol dm^{-3} terephthalate) under argon in the sonication bath directly (inset) and in an immersed small probing vessel (at two different positions). The solid lines represent the results for the axial position, the dotted lines for an extra-axial position (for details of the setup see [40]), as a function of the power input at the transducer (frequency: 321 kHz).

of gas. We conclude that in the lower power range the ultrasound happens to be axially "focused" to some extent under these conditions (of course, the presence of the probe itself influences the sound field). This observation is in support of the contention made above that the sonic field is inhomogeneous.

3.2 Operating Gas

The OH radical yield depends on the nature of the operating gas. The temperature reached in the compressed bubble depends mainly on the ratio of specific heats c_p/c_v of the gas and its thermal conductivity. Of course, the presence of water vapor and of other polyatomic volatile components of the solution will depress the value of c_p/c_v below that of a pure monoatomic or diatomic gas. Any endothermic reaction will further depress this ratio as it acts to increase the apparent heat capacity. In the perfectly adiabatic limit, T_{max} and V_{min} of the collapsed bubble are linked by the expression $T_{max}/T_0 = (V_0/V_{min})^{c_p/c_v - 1}$, wherefrom it is seen that in a situation where c_p/c_v approaches unity, the temperature rise must be small even at high compression ratios. Further, the higher the thermal conductivity is, the more efficient the heat transfer out of the compressed bubble. The ratio c_p/c_v is highest for the noble gases, but in helium, the latter effect is especially important (see [29, p. 331]) so that under this noble gas, the OH radical yield is 0.5 times lower than under argon. The trend of increasing OH radical production with decreasing heat conductivity in the noble gases is well established [73].

The hydrogen peroxide and OH radical yields of argon–helium mixtures are shown in Figure 8 (inset). As the helium content increases from zero, there is a rise in hydrogen peroxide production, which may be explained on the basis of an increased cooling of the gaseous fringes of the bubble by thermodiffusion (i.e., selective loss of hot helium atoms from this zone), so that any hydrogen peroxide formed in this zone is better preserved. As the helium content increases further, the bubble core where the cracking of water molecules mainly occurs increasingly experiences heat loss, which suppresses the production of OH radicals, leading to the eventual decline of the curves.

In argon–hydrogen mixtures, both the OH and H_2O_2 yields decline more steeply with increasing H_2 content than in the argon–helium case (Figure 8) for three reasons: consumption of OH radicals by way of the reverse of reaction (4), a higher thermal conductivity, a lower ratio c_p/c_v. Moreover, the production of H_2O_2 is especially strongly suppressed because of reaction (9), so that its ranking with respect to the OH radical yield is the inverse of that obtained in the Ar/He situation.

In contrast to the noble gases, diatomic and especially polyatomic operating gases may influence the sonolysis of a solute by their own thermal decomposition into more or less reactive fragments. Among the diatomic operating gases, N_2 is essentially stable up to 3000 K but O_2 begins to show dissociation into the atoms already near 2500 K (about 1%) [74,75]. More importantly, O_2 may react with solute free radicals, thus leading to the formation of peroxyl radicals in the solution

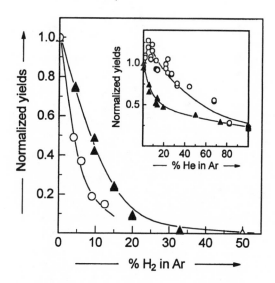

Figure 8. Normalized yields of free OH radicals (▲) and H_2O_2 (○) in the sonolysis of aqueous terephthalate solutions (2×10^{-3} mol dm^{-3}) of Ar/H$_2$ and Ar/He (inset) gas mixtures.

and combustion processes in the gas phase. In the sonolysis of aqueous solutions, peroxyl radical formation is of necessity always a concern since the thermolysis of water vapor generates some O_2 [reactions (5), (8), and (10)].

During the initial degassing stage in the sonolysis of gas-saturated solutions, cavitational action is reduced because of the weakening of the sonic field due to a high degree of effervescence [see [29, p. 163]). After partial degassing, OH radical production takes place more effectively (Figure 9) [40]. This phenomenon is clearly observed only in the lower power range, possibly because at higher powers the rate of degassing is sufficiently fast on the time scale of the sonolysis experiment.

3.3 Temperature and Pressure

There is a dramatic effect of temperature on the yield of free OH radicals, which occurs in parallel with the H_2O_2 yield in the absence of any scavenger (Figure 10) [76].

This falloff (see [77]) with rising temperature of the liquid has several causes. The degree of heating of the bubble contents will be less since the proportion of water vapor increases with increasing temperature (e.g., ≤ 1% at ≤ 10 °C and 30% at 70 °C) and therefore the value of c_p/c_v of the gas–vapor mixture decreases. Further, the component of the compressive force that is due to the surface tension decreases. Moreover, it is easy to see that as the boiling point is approached, bubble

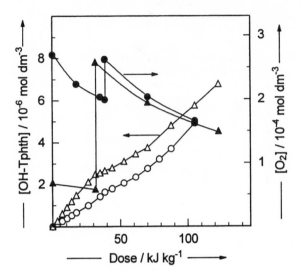

Figure 9. Sonolysis of aqueous terephthalate $(2 \times 10^{-3} \text{ mol dm}^{-3})$ dosimeter solutions in the presence of air. Formation of 2-hydroxyterephthalate and loss of oxygen through sonic degassing. Sonication was carried out at a frequency of 585 kHz. Oxygen loss curves: (●) initial air-saturation at atmospheric pressure with resaturation after a dose of 40 kJ kg^{-1}; (▲) solution briefly subjected to a water pump vacuum before sonication, with resaturation after a dose of 33 kJ kg^{-1}. Hydroxyterephthalate yield curves (○, △): concentration increase depending on the gas status of the solutions. Curves with circular and triangular symbols are in respective correspondence.

compression must cease because the pressure of the bubble contents balances the external pressure which is the other component of the compressive force. It has been shown recently that the phenomenon of sonoluminescence, which has been attributed largely to an emission of electronically excited OH radicals, is strongly dependent on the bulk temperature of the water [76]. On going from 11 to 40 °C, sonoluminescence is reduced by 90% at a frequency of 337 kHz. Figure 10 shows a decrease in the free radical yield of only 50% over this temperature range. This is compatible with the higher percentage drop in sonoluminescence activity; as the overall activation energy of the process leading to electronic excitation is higher than that of OH radical formation, the effect of lowering the temperature on bubble compression must act more strongly on the sonoluminescence yield.

A similar falloff behavior is observed when, at room temperature, sonolysis is carried out under reduced external pressure. In this case, sonochemical activity effectively ceases below about 0.1 atm (Figure 11). Reducing the external pressure makes for a lower incidence of cavitation events, as well as providing a lesser compressive force.

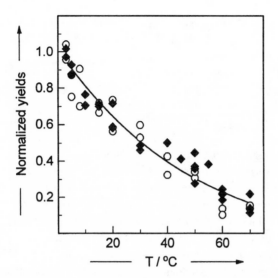

Figure 10. Sonolysis of Ar-saturated water at 321 kHz and an intensity of 170 W kg^{-1}. Normalized yields of H_2O_2 (○, no additive) and OH radicals (◆, determined with the terephthalate dosimeter, 2×10^{-3} mol dm^{-3} in the presence of 2×10^{-4} mol dm^{-3} IrCl$_6^{2-}$) as a function of the bath temperature.

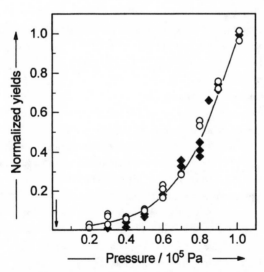

Figure 11. Sonolysis of Ar-saturated water at 321 kHz and an intensity of 170 W kg^{-1} at a temperature of 20 °C. Normalized yields of H_2O_2 (○, no additive) and OH radicals (◆, determined with the terephthalate dosimeter, 2×10^{-3} mol dm^{-3} in the presence of 2×10^{-4} mol dm^{-3} IrCl$_6^{2-}$) as a function of the external pressure. The arrow indicates the vapor pressure of water at this temperature (2.3×10^3 Pa).

4. THE OH RADICAL AND THE DECOMPOSITION OF VOLATILE SOLUTES

It is known that in sonolysis, the effectivity scale of various OH radical scavengers in suppressing the formation of H_2O_2 [40,78] differs greatly from that observed in radiation chemistry where the effectivity scale is in fact represented by the rate constants of the reaction of these compounds with the OH radical. To explain this difference, it has been suggested that scavengers accumulate at the bubble–liquid interface according to their degree of hydrophobicity [1,78], where they intercept the OH radicals formed on cavitation. Because of the remarkably large scale of this enhancement effect, we have previously put forward a suggestion that provides a link between the degree of suppression of the hydrogen peroxide yield and the vapor pressure of the solute in the bubble [79], which is determined by its fugacity in the solution. The partial pressure will be enhanced depending on the extent of thermal ablation of the boundary layer hydrophobically enriched in the volatile solute. It has been estimated that the boundary layer is heated to supercritical temperatures to a depth of 200 to 500 nm [23], which suggests that some thermal ablation should occur.

In the present paper, we expand on the hypothesis of gas-phase OH radical scavenging and provide a hypothesis that complements the approach based on hydrophobicity [78]. In this picture, OH radicals may already have been scavenged at their source, within the bubble. On account of its high reactivity, as well as the elevated temperature within the compressed bubble, reaction will in general have occurred after only a few nonproductive encounters, i.e., the gas-phase decomposition of a volatile solute has a OH-radical-induced as well as a pyrolytic (see [3]) component.

The following simple argument demonstrates that the OH radical should in fact have a good chance to experience a reactive encounter inside the bubble even at low solute concentrations. We assume that the volatile solute enters the bubble from the solution only by diffusion during expansion. Consider a bubble with an initial radius r of 5×10^{-3} cm (see [25]), initial pressure p_0 about 0.5 atm ($= 0.5 \times 10^5$ Pa) at room temperature, that is about to undergo cavitation. Let the solution contain 10^{-3} mol dm^{-3} t-butanol, which corresponds to a partial pressure $p_{\text{t-BuOH}}$ of 1.2 Pa [80]. We wish to know the ratio d/r where d is the average distance between the point of encounter with a t-butanol molecule and the point of departure of the OH radical within the bubble. A value of this ratio below unity implies appreciable OH radical scavenging. We may consider the situation as it presents itself at the beginning of the cavitation event since d/r is expected to remain essentially unchanged on compression of the bubble; d is given by the Einstein–Smoluchowski equation [where D is the diffusion coefficient, which is about 0.2 cm^2 s^{-1}, see [81]]:

$$d = (2 D t)^{1/2} \tag{39}$$

The time t elapsed between two encounters is given by the following expression, where v is the velocity of the OH radical (600 m s^{-1}) and l the length of the path traced up to the next encounter with a t-butanol molecule:

$$t = \frac{l}{v} \qquad (40)$$

The latter is given by

$$l = \frac{p}{p_{t\text{-BuOH}}} \lambda \qquad (41)$$

where the mean free path λ is about 2×10^{-5} cm at the conditions assumed.

After insertion of the numbers, one obtains an estimated value of 0.47 for r/d. While at room temperature about one in ten encounters is productive ($k = 6 \times 10^8$ dm^3 mol^{-1} s^{-1}), the productivity ratio becomes more favorable at high temperatures within the bubble. The odds improve when the initial radius of the bubble or the scavenger concentration in the solution is larger, or if some thermal ablation takes place. They further improve as the t-butanol in the bubble suffers thermal decomposition, insofar as each t-butanol-derived radical or fragmentation product (e.g., isobutene) is itself an even better OH radical scavenger.

The hypothesis that the OH radical scavenging effectivity of organic solutes is related to a tendency of accumulating in the boundary layer of the bubble deserves some further comment. In order to assess the importance of the contribution of the solute in the boundary layer to the scavenging of OH radicals as reflected in the reduction of the H_2O_2 yield, it is necessary to form an idea regarding the enhancement of the concentration of the solute in the boundary layer compared with that in the bulk. This is attempted in the following manner. The concentration of a solute in the gas phase c_g above its dilute aqueous solution of concentration c_w is given by expression (42) [80] (this reference contains fugacity data, characterized by γ, for various solutes), and corresponds to a partial pressure p according to expression (43):

$$c_g = \frac{c_w}{\gamma} \qquad (42)$$

$$p = c_g RT \qquad (43)$$

Ideally in the hypothetical state of the absence of hydrophobic or hydrophilic interaction, the vapor pressure of a solute is proportional to its molar fraction x in the surface layer of the solution [expression (44)]. On the other hand, in dilute solution x is given in terms of concentrations by expression (45) where c represents the solute concentration in the gas/liquid boundary layer and c_{H_2O} equals 55.5 mol dm^{-3}.

$$\frac{p}{p_{neat}} = x \tag{44}$$

$$x = \frac{c}{c_{H_2O}} \tag{45}$$

The vapor pressure of the neat solute p_{neat} at temperature t (in degrees C) can be represented by the Antoine equation [expression (46)] where A, B, and C are constants that are specific for a solute [82]:

$$\log p_{neat} = A - \frac{B}{(t + C)} \tag{46}$$

Combining the foregoing expressions and solving for c, we obtain

$$c = c_g \frac{RTc_{H_2O}}{p_{neat}} \tag{47}$$

On substitution of the quantities for the case of t-butanol ($\log \gamma = 3.31$, $c_w = 1 \times 10^{-3}$ mol dm^{-3}, $T = 298$ K, $A = 7.23$, $B = 1107$, $C = 172$), we obtain $c = 1.7 \times 10^{-2}$ mol dm^{-3}, which is taken as an upper limit to the concentration of t-butanol in the boundary layer of the bubble before it enters its compressive phase. This implies a hydrophobic-enhancement factor of 17. A t-butanol concentration of 1.7×10^{-2} mol dm^{-3} represents a scavenging capacity of 1.1×10^7 s^{-1}. We have seen above that 7×10^{-3} mol dm^{-3} iodide ion ([49] reports 5×10^{-3} mol dm^{-3}) is needed, representing a scavenging capacity of 7.7×10^7 s^{-1}, to cut the H_2O_2 yield in half (Figure 2). To achieve the same effect under purely solution-kinetics conditions with t-butanol, a concentration of 0.13 mol dm^{-3} would be necessary. As a boundary-layer concentration, this corresponds to a bulk concentration of 7.6×10^{-3} mol dm^{-3}, which is far beyond the t-butanol bulk concentration actually needed (near 0.2×10^{-3} mol dm^{-3}, see Figure 12) to reduce the H_2O_2 yield by half. This means that in the case of t-butanol, hydrophobic enrichment of the boundary layer cannot serve as the sole explanation for the enhancement of the OH radical scavenging effect, compared to what is known from radiation chemistry.

It is feasible from the iodide (Figures 2 and 3, and [49]) and t-butanol data (Figure 12) to extract further, more precise information about the concentration and fate of the OH radical in the boundary layer by means of the formalized kinetic approach laid out in the Appendix (loss of OH radicals out of this layer by diffusion has been neglected). As an approximation, the concentration of the species involved is considered not to vary across the layer. The concentrations of the OH radical scavengers have been assumed to remain constant.

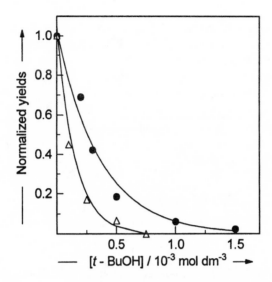

Figure 12. Suppression of the proportion of sonolytically generated OH radicals (Ar-saturated aqueous solution, 321 kHz, 170 W kg^{-1}) at various concentrations of t-butanol. Normalized yields of free OH radicals (●, determined with 2×10^{-3} mol dm^{-3} terephthalate in the presence of 2×10^{-4} mol dm^{-3} IrCl$_6^{2-}$) and H$_2$O$_2$ (Δ, in the presence 2×10^{-3} mol dm^{-3} terephthalate) in aqueous solutions of t-butanol.

We begin with the determination of the OH radical concentration where the sole scavenger is iodide. With the assumption that the nonvolatile scavenger iodide does not interfere with the OH radical in the interior of the bubble, we may calculate the maximum (in the absence of scavenging in the gas phase) achievable concentration x_0 (symbols and numbering are those used in the Appendix) of OH in the boundary layer, making use of

$$\frac{[\text{H}_2\text{O}_2]_\alpha}{[\text{H}_2\text{O}_2]_{\alpha_0}} = 1 - \frac{\alpha}{x_0} \ln\left(1 + \frac{x_0}{\alpha}\right) \tag{48}$$

This equation corresponds to Eq. (A16) of the Appendix, with the simplification that $x = x_0$ because of the absence of gas-phase OH radical scavenging; $[\text{H}_2\text{O}_2]_{\alpha_0}$ is the H$_2$O$_2$ concentration in the boundary layer in the absence of any scavenger ($\alpha_0 = 0$). Moreover, in the expression for α [Eq. (A6)], we replace [T] by [I$^-$] (no second scavenger, i.e., [S] = 0), so that with the iodide concentration needed to cut the H$_2$O$_2$ yield in half (about 7×10^{-3} mol dm^{-3}, Figures 2 and 3), $\alpha = k_2\,[\text{I}^-]_{1/2}/2k_1$, where $k_1(\cdot\text{OH} + \cdot\text{OH}) = 5.5 \times 10^9$ dm^3 mol^{-1} s^{-1} and $k_2(\cdot\text{OH} + \text{I}^-) = 1.1 \times 10^{10}$ dm^3 mol^{-1} s^{-1} [66]. Inserting the values for these quantities and $[\text{H}_2\text{O}_2]_\alpha/[\text{H}_2\text{O}_2]_{\alpha_0} = 0.5$, it can be shown that this equation is fulfilled by $x_0 = 1.2 \times 10^{-2}$ mol dm^{-3}. This value for the maximal OH radical concentration in the boundary layer is moderately higher but still roughly similar to an estimate of 5×10^{-3} mol dm^{-3} determined under different frequency conditions [49].

Now to the t-butanol data (Figure 12). Here, $\alpha_0 = k_2[T]/2k_1$ and $\alpha = (k_2[T] + k_3[S])/2k_1$ ([T] = terephthalate concentration = 2×10^{-3} dm^3 mol^{-1} s^{-1}, S stands for t-butanol; $k_2 = 3.3 \times 10^9$ dm^3 mol^{-1} s^{-1} [60], $k_3 = 6 \times 10^8$ dm^3 mol^{-1} s^{-1} [66]). We anticipate that in the presence of t-butanol the initial concentration of OH radicals is reduced from x_0 to x by gas-phase scavenging. This is reflected in the following expression [see Eq. (A17)]:

$$\frac{[P]_\alpha}{[P]_{\alpha_0}} = \frac{\ln\left(1 + \dfrac{x}{\alpha}\right)}{\ln\left(1 + \dfrac{x_0}{\alpha_0}\right)} \tag{49}$$

For [S] we take the product of the bulk concentration, read off the abscissa in Figure 12, times the hydrophobic-enhancement factor of 17 (see above). At a bulk concentration of 2.5×10^{-4} mol dm^{-3}, we read off $[P]_\alpha/[P]_{\alpha_0} = 0.5$ in Figure 12 and insert this in Eq. (49). Solving Eq. (49), we obtain $x/\alpha = 3.6$, from which $x = 3.0 \times 10^{-3}$ mol dm^{-3} is calculated.

Is this result compatible with the H_2O_2 data? We substitute $x/\alpha = 3.6$ into Eq. (50) [see Eq. (A16)]. For 2.5×10^{-4} mol dm^{-3} t-butanol, $[H_2O_2]_\alpha/[H_2O_2]_{\alpha_0} = 0.2$ is read off in Figure 12.

$$\frac{[H_2O_2]_\alpha}{[H_2O_2]_{\alpha_0}} = \frac{x - \alpha \ln\left(1 + \dfrac{x}{\alpha}\right)}{x_0 - \alpha_0 \ln\left(1 + \dfrac{x_0}{\alpha_0}\right)} \tag{50}$$

On substitution, the right-hand side of Eq. (50) is calculated at 0.17, versus 0.20 to the left. This shows that $x = 3.0 \times 10^{-3}$ mol dm^{-3} and $x/\alpha = 3.6$ essentially fulfill Eq. (50).

Table 1 shows how the proportion of OH radicals x/x_0 that reach the boundary layer diminishes with increasing t-butanol concentration. The shortcoming of the above kinetic model, i.e., that the substrate concentrations are assumed to remain constant, turns out to be less serious than would appear at first sight since the OH-radical-reaction and subsequent products act as OH radical scavengers as well. Moreover, it will in no way detract from the conclusion that gas-phase OH radical scavenging is important. While the hydroxyterephthalate yields will tend to appear smaller than would be the case in the absence of secondary reactions, this effect will decrease as the concentration of the second scavenger increases, with the consequence that the ratio $[P]_\alpha/[P]_{\alpha_0}$ [see expression (49)] observed decreases less steeply with increasing [t-BuOH] than it ideally would in the absence of the second scavenger.

This kinetic model assumes that there is no diffusive loss of OH radicals from the reaction zone, i.e., the t-butanol-enriched boundary layer. This cannot be strictly true. However, in order to develop an idea as to the possible extent of diffusive loss, we shall again make use of the Einstein–Smoluchowski equation [see Eq. (39)]. For the *aqueous* solution situation, we take for the OH radical $D = 2 \times 10^{-5}$ cm^2 s^{-1} (see [81]), $v = 600$ m s^{-1}, the mean free path $\lambda = 0.1$ nm, [t-BuOH] $= 1 \times 10^{-3}$ mol dm^{-3} (in the boundary layer), and $c_{H_2O} = 55.5$ mol dm^{-3}. Since at room temperature, about one encounter in ten leads to OH radical consumption, we estimate $l = (55.5/1 \times 10^{-3}) \times 10^{-8} \times 10 = 5.5 \times 10^{-3}$ cm. Insertion of the quantities into Eq. (39) gives a value of 1.9×10^{-6} cm for d, which implies that at [t-BuOH] $= 1 \times 10^{-3}$ mol dm^{-3} and above, most of the OH radicals will be consumed inside a boundary layer with a thickness of several tens of nanometers only, and OH radical reactions in the bulk will play at best a minor role. In a situation where [·OH] exceeds the scavenger concentration in the boundary layer, the scavenger may be overwhelmed, and OH radicals will penetrate somewhat deeper into the solution. In this situation it is possible that some of the OH radicals also react with scavenger radicals. In a favorable case such as the iodide system, this can lead to the same result as if the scavenger were not partly depleted (·OH + I· → HOI; HOI + I$^-$ → I$_2$ + OH$^-$).

The foregoing provides a strong indication that the data in Figure 11 and the conceptual approach to their explanation are consistent.

While compounds such as t-butanol appear to exhibit an inordinately large scavenging power toward OH radicals in the liquid phase if the gas-phase processes are not taken into account, the situation is the reverse at the other end of the hydrophobicity scale [78]. Formic acid reacts with the OH radical at 1.3×10^8 dm^3 mol^{-1} s^{-1} [66], i.e., about 80 times more slowly than iodide. But about 1.4 mol dm^{-3} formic acid, i.e., a concentration about 300 times greater than for iodide, is needed to achieve an equal degree of H$_2$O$_2$ suppression. The discrepancy between observation and expectation is even greater when it is remembered that at this concentration about one-tenth of the formic acid (p$K_a = 3.75$) is dissociated, that the

Table 1. The Variation of the Proportion x/x_0 of OH Radicals that Escape from the Bubble[a]

c_w (10^{-3} mol dm^{-3})	c (10^{-3} mol dm^{-3})	x (10^{-3} mol dm^{-3})	x/x_0
0	0	12	1
0.25	4.3	3	0.25
0.5	8.5	1.2	0.1
1.0	17	0.5	0.04

Notes: [a]c_w, t-butanol concentration in the bulk; c, t-butanol concentration in the boundary layer; x, initial OH radical concentration in the boundary layer. $c = 17\, c_w$ (see text).

formate ion reacts about 20 times as fast with the OH radical, and that some OH radical scavenging must already occur in the gas phase on account of formic acid volatility. We think that sonolytically produced O_2 [in reaction (9)] reacts with the formate radical CO_2^- under the formation of superoxide HO_2^-/O_2^- which furnishes H_2O_2 on disproportionation, in partial compensation of the shortfall of H_2O_2 caused by OH radical scavenging.

4.1 t-Butanol

It has been shown that at a concentration of 10^{-3} mol dm^{-3} of this solute, the quantity of OH radicals reaching the liquid phase is not more than about 4% of that observed in the absence of this solute (Table 1). In the absence of a volatile scavenger, $G(OH) = 16 \times 10^{-10}$ mol J^{-1}. We therefore expect a G value of about 0.6×10^{-10} mol J^{-1} for the OH-induced decomposition of t-butanol in the aqueous phase. Given that under these conditions the dominant product is 2,5-dimethylhexane-2,5-diol, formed with the stoichiometry implied by reactions (51) and (52), a G value of 0.3×10^{-10} mol J^{-1} would be expected for this diol.

$$
\begin{array}{ccccc}
\underset{\underset{CH_3}{|}}{\overset{\overset{CH_3}{|}}{CH_3\text{-}C\text{-}OH}} & \xrightarrow[\quad(51)\quad]{\overset{\cdot OH\,/\,-H_2O}{H\cdot/\,-H_2}} & \underset{\underset{CH_3}{|}}{\overset{\overset{CH_3}{|}}{\cdot CH_2\text{-}C\text{-}OH}} & \xrightarrow[\quad(52)\quad]{2\,x} & \underset{\underset{CH_3}{|}}{\overset{\overset{OH}{|}}{CH_3\text{-}C}}\text{-}CH_2\text{-}CH_2\text{-}\underset{\underset{CH_3}{|}}{\overset{\overset{OH}{|}}{C\text{-}CH_3}}
\end{array}
$$

However, the G value observed for this product is only about 0.03×10^{-10} mol J^{-1} [83,84], i.e., about one-tenth of what is expected. This suggests that the proportion of OH radicals entering the aqueous phase is even smaller than estimated, or that precursor radicals are lost in combination with other radicals that are produced in the pyrolytic reactions and then enter the aqueous phase.

The fact that traces of this diol are also observed in the sonolysis of more highly concentrated t-butanol solutions where essentially complete scavenging of OH radicals within the bubble is expected, suggests that some of the diol or its precursor are formed in the gas phase as well.

The sonolysis of t-butanol gives rise to a great variety of products, among them a considerable number of hydrocarbons [83,84]. The results lead to the conclusion that at the concentrations studied, almost all of the decomposition of the t-butanol takes place within the bubble. The products are mostly formed in thermolytic reactions (see [85]). At low t-butanol concentrations, their sonolytic yields increase roughly in proportion with the concentration (see Figure 13, inset).

At higher t-butanol concentrations, the yields pass through a maximum and then reach again low values at concentrations above 0.2 molar. Figure 13 shows the behavior of two examples, isobutene and methane. This is explained as follows. At first, the amount of pyrolysis products increases with the partial pressure of the substrate. When the partial pressure reaches a level so as to begin to affect the thermal characteristics of the bubble contents through a lowering of c_p/c_v, or when the substrate concentration in the liquid becomes high enough to perhaps affect the intensity of cavitation (e.g., by reducing the surface tension), the temperature reached inside the bubble is lower, leading to a drop in product formation, as well as a change in product composition. It is noted that isobutene becomes *relatively* more prominent at the higher *t*-butanol concentration (Figure 13). This is in keeping with the observation that the shock-tube pyrolysis of *t*-butanol below 1300 K proceeds via the nonradical pathway (53), yielding water and isobutene, while a homolytic process begins to contribute above this temperature [reaction (54)] [85].

Here it is important to recall that radicals undergo much more facile fragmentation than their parent molecules, and fragmentations such as reactions (56)/(57) must be expected, partly contributing to the H_2 yield, and reactions (60) and (61) would contribute to the formation of CO, which is a prominent product [83,84].

An interesting feature is shown in Figure 14. The H_2 yield rises steeply at very small scavenger concentrations from the value of $\approx 8 \times 10^{-10}$ mol J^{-1} observed in the absence of *t*-butanol, in good chemical-balance agreement with $G(H_2O_2) = 7.5 \times 10^{-10}$ mol J^{-1} (see above), to peak well before methane does. The explanation

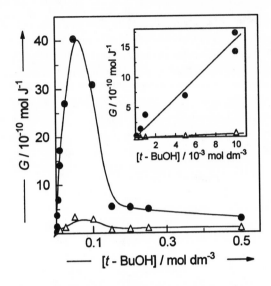

Figure 13. Sonolysis of Ar-saturated aqueous solutions of *t*-butanol (321 kHz, 170 W kg^{-1}): yield of methane (●) and isobutene (Δ) depending on the concentration of *t*-butanol. Inset: yield of these products at low concentrations of *t*-butanol.

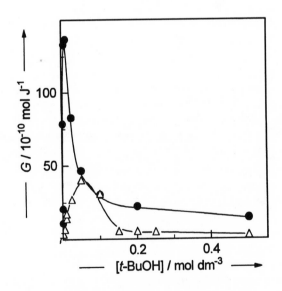

Figure 14. Sonolysis of Ar-saturated aqueous solutions of *t*-butanol (321 kHz, 170 W kg^{-1}): yield of H$_2$ (●) and methane (Δ) depending on the concentration of *t*-butanol.

considers the processes in the gas phase and is as follows. Hydrogen atom abstraction by H· from the substrate [reaction (55)] competes on much more favorable terms with the recombination reactions (6) and (7) than does the analogous reaction with H_2O [reaction (4)]. As the substrate concentration rises, free radical formation by pyrolysis increases in importance [e.g., reaction (54)]. This provides an even more favorable channel for H atom disappearance, i.e.,

$$H· + ·CH_3 \rightarrow CH_4 \tag{63}$$

Accordingly, the H_2 yield must drop while that of other products, e.g., methane, still rises.

A general flattening out of the yields (see Figures 13 and 14) beyond a *t*-butanol concentration of about 0.2 molar [83,84] indicates that from this concentration onward, the properties of the boundary layer, including the solute fugacity, begin to resemble those of neat *t*-butanol (10 molar): 0.2 molar times the hydrophobicity factor of 17 (see above) is not far from 10 molar!

4.2 4-Nitrophenol

The contribution of thermolytic decomposition in sonolysis is not restricted to solutes with relatively low boiling points such as *t*-butanol. 4-Nitrophenol provides an example of a higher-boiling compound whose partial pressure over its 5×10^{-3} molar acidified solution at 20 °C is 1.7×10^{-2} Pa [86]. In basic solution where the ionic form predominates ($pK_a = 7.1$), the solute is no longer volatile, thermolytic products are absent, and OH-radical-induced chemistry determines the product distribution (Figure 15) [79,83,88]. This is a general phenomenon, also observed in, e.g., 4-chlorophenol [89].

Remarkably, the concentration dependence of the H_2O_2 yield ([83], ○ in Figure 16) is quantitatively similar to that shown by *t*-butanol (Figure 12), despite the large difference between the gas-phase partial pressures at equal concentrations c_w (see above). If this falloff is due to gas-phase scavenging, it would mean that the volatile-substrate content in the bubble is not just determined by solute fugacity but that part of the substrate-enriched boundary layer is thermally ablated. From the solubility of 4-nitrophenol (8.3×10^{-2} mol dm^{-3} [86], where equilibrium exists with the neat compound, which is about 10 mol dm^{-3}) a high hydrophobic-enrichment factor (about 80) is estimated. (Enrichment of the nitrophenolate anion will *not* take place; therefore, one expects a lesser decline of the H_2O_2 yield under alkaline conditions.) This is indeed the case (● in Figure 16), the aspect at pH 10 being very similar to the terephthalate situation.

Figure 17 shows that 4-nitrophenol consumption is considerably higher at pH 4, while at pH 10 the picture closely resembles the behavior of the ionic substrates, i.e., terephthalate (see above).

Alternatively, the steep decline of $G(H_2O_2)$ at pH 4 with increasing 4-nitrophenol concentration could well be largely due to OH radical scavenging in the boundary layer because of the high enrichment factor, in contrast to the *t*-butanol case where

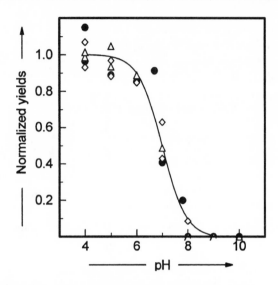

Figure 15. Sonolysis of Ar-saturated aqueous solutions of 4-nitrophenol (5×10^{-3} mol dm^{-3}, 321 kHz, 170 W kg^{-1}). pH dependence of the yields of phenol (●) and gaseous products (◇: acetylene and △: ethene).

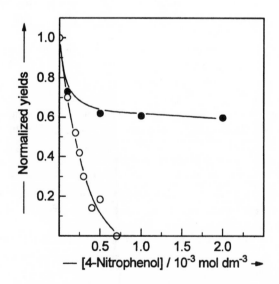

Figure 16. Suppression of H_2O_2 production in Ar-saturated aqueous solutions pH 4 (○) and pH 10 (●) at various concentrations of 4-nitrophenol. The H_2O_2 yields have been normalized to zero 4-nitrophenol concentration (321 kHz, 170 W kg^{-1}).

the volatility is much higher and the enrichment factor much lower. The steady increase of the nitrocatechol yield ([83], Figure 18) supports this view. If there were an important degree of gas-phase scavenging, one would expect the formation of this product to be suppressed at high substrate concentrations, just like 2,5-dimethylhexane-2,5-diol is suppressed in the *t*-butanol case [79,83,84]. The monitoring of OH radicals by means of terephthalate proved unfeasible in the 4-nitrophenol system.

The dependence of the thermolysis product phenol on the substrate concentration [83] is shown in the inset of Figure 18. Its linear increase, taken together with the increase of the nitrocatechol, suggests that the source of the former does not lie in the gas phase, unless the latter were also a gas-phase product. It is more likely that nitrocatechol is formed in the liquid phase (boundary layer), following OH radical addition to the nitrophenol, in competition with H_2O_2 formation (i.e., as H_2O_2 goes down, nitrocatechol goes up). But a substantial production of phenol in the gas phase would imply a corresponding gas-phase presence of nitrophenol, increasingly sequestering OH radicals and so forcing an early downturn of the nitrocatechol yield. This implies that thermal decomposition of 4-nitrophenol in the boundary layer, apart from its OH-radical-induced destruction, including the formation of phenol by aquathermolysis [87], is a strong possibility, cf. also ref. 90.

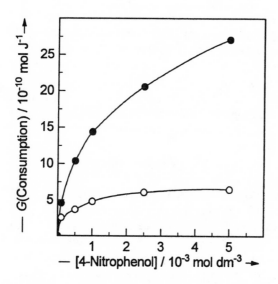

Figure 17. Consumption of 4-nitrophenol at pH 4 (●) and pH 10 (○) during the sonolysis of Ar-saturated solutions as a function of its initial concentration (321 kHz, 170 W kg^{-1}).

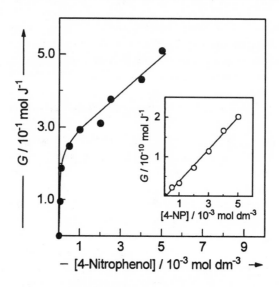

Figure 18. Sonolysis of Ar-saturated solutions of 4-nitrophenol at pH 4 (321 kHz, 170 W kg^{-1}). Yield of 4-nitrocatechol (●) as a function of the 4-nitrophenol concentration. Inset: yield of phenol (○) as a function of the 4-nitrophenol concentration

The discontinuity in the rise of G(nitrocatechol) near 2×10^{-4} mol dm^{-3} nitrophenol (Figure 18) suggests that beyond this concentration (which corresponds to a concentration in the boundary layer of $\approx 2 \times 10^{-2}$ mol dm^{-3} assuming a hydrophobic enhancement factor of ~ 100; this value of 2×10^{-2} is close to the maximum achievable concentration of OH radicals estimated at 1.2×10^{-2} mol dm^{-3}, see above), depletion of OH radicals slows down the initial fast rise. In contrast, the production of phenol, which is not linked to OH radical action, shows a linear increase as expected (Figure 18, inset).

5. CONCLUSIONS

The chemical effects of ultrasound in aqueous solution resemble those of ionizing radiation in that OH and other free radicals are produced in both cases. Hydroxyl radical reactions, e.g., with terephthalate or iodide, may be used for sonic-field dosimetry. The ultrasonic cavitation process creates a hot gas-phase and a gas–liquid boundary layer, which represent physicochemical environments that do not exist in radiation chemistry. With solutes that are neither volatile nor hydropho-bic (e.g., electrolytes), the resemblance to radiation chemistry is relatively close; the sonochemistry of such compounds is characterized by OH-radical-induced reactions in the liquid phase. In the case of volatile compounds, aqueous-phase OH radical chemistry is at the same time overshadowed by pyrolytic reactions of the solute and suppressed by gas-phase OH radical scavenging within the cavitational bubble. Hydrophobicity causes an enrichment of the solute in the boundary layer, which further compounds the strongly nonhomogeneous nature of sonochemistry.

6. APPENDIX

We consider the competing reactions

$$\cdot OH + \cdot OH \overset{1}{\rightarrow} H_2O_2 \tag{A1}$$

$$\cdot OH + T \overset{2}{\rightarrow} P \tag{A2}$$

$$\cdot OH + S \overset{3}{\rightarrow} Q \tag{A3}$$

(e.g., T = terephthalate, S = t-butanol, P = hydroxyterephthalate). The time dependence of the OH radical concentration is described by

$$\frac{d[OH]}{dt} = -2\,k_1[OH]^2 - (k_2[T] + k_3[S])\,[OH] \tag{A4}$$

Renaming for convenience

$$[OH] = x(t) \tag{A5}$$

$$\frac{k_2[T] + k_3[S]}{2\,k_1} = \alpha \tag{A6}$$

$$2\,k_1 = \beta \tag{A7}$$

one obtains after integration

$$\ln \frac{x(t)(x_0 + \alpha)}{x_0(x(t) + \alpha)} = -\alpha\beta t \tag{A8}$$

or

$$x(t) = \frac{x_0\,\alpha\,e^{-\alpha\beta t}}{(x_0 + \alpha - x_0 e^{-\alpha\beta t})} \tag{A9}$$

where x_0 represents the OH radical concentration at $t = 0$. The buildup of the products H_2O_2 and P (these are monitored) is described by

$$\frac{d[H_2O_2]}{dt} = k_1(x(t))^2 \tag{A10}$$

$$\frac{d[P]}{dt} = k_2[T]\,x(t) \tag{A11}$$

After integration and introduction of the boundary conditions $[H_2O_2]_{t=0} = 0$ and $[P]_{t=0} = 0$, one obtains expressions (A12) and (A13) for $[H_2O_2]_{t=\infty}$ and $[P]_{t=\infty}$ i.e., after all of the OH radicals have reacted.

$$[H_2O_2]_{t=\infty} = \frac{1}{2}\left(x_0 - \alpha \ln \frac{x_0 + \alpha}{\alpha}\right) \tag{A12}$$

$$[P]_{t=\infty} = \frac{k_2[T]}{\beta} \ln \frac{x_0 + \alpha}{\alpha} \tag{A13}$$

By extension:

$$[Q]_{t=\infty} = \frac{k_3[S]}{\beta} \ln\left(\frac{x_0 + \alpha}{\alpha}\right) \tag{A14}$$

For the meaning of Q, see expression (A3). Inspection shows that

$$2\,[H_2O_2]_{t=\infty} + [P]_{t=\infty} + [Q]_{t=\infty} = x_0 \tag{A15}$$

as required.

These quantities are dependent on the concentration of the scavengers T and S. The normalization of these quantities obtained in the presence of a second scavenger, characterized by α, with respect to the quantities obtained in the *absence* of the second scavenger, characterized by $\alpha = \alpha_0$, leads to expressions (A16) and (A17), where it must be borne in mind that the initial concentration of OH radicals may be diminished, say from x_0 to x, if the scavenger S is volatile and reacts with OH radicals already in the gas phase before they have entered the liquid phase.

$$\frac{[H_2O_2]_\alpha}{[H_2O_2]_{\alpha_0}} = \frac{x - \alpha \ln\left(1 + \dfrac{x}{\alpha}\right)}{x_0 - \alpha_0 \ln\left(1 + \dfrac{x_0}{\alpha_0}\right)} \tag{A16}$$

$$\frac{[P]_\alpha}{[P]_{\alpha_0}} = \frac{\ln\left(1 + \dfrac{x}{\alpha}\right)}{\ln\left(1 + \dfrac{x_0}{\alpha_0}\right)} \tag{A17}$$

REFERENCES

[1] Henglein, A. *Ultrasonics*, 25 (1987) 6.
[2] Suslick, K. S. *Science*, 247 (1990) 1439.
[3] Henglein, A. *Adv. Sonochem.*, 3 (1993) 17.
[4] Weissler, A. *J. Am. Chem. Soc.*, 81 (1959) 1077.
[5] Makino, K., Mossoba, M. M., and Riesz, P. *J. Phys. Chem.*, 87 (1983) 1369.
[6] Makino, K., Mossoba, M. M., and Riesz, P. *Radiat. Res.*, 96 (1983) 416.
[7] Makino, K., Mossoba, M. M., and Riesz, P. *J. Am. Chem. Soc.*, 104 (1982) 3537.
[8] Riesz, P., Berdahl, D., and Christman, C. L. *Environ. Health Perspect.*, 64 (1985) 233.
[9] Gmelins Handbuch der Anorganischen Chemie. *Sauerstoff*, 8th ed. Verlag Chemie, Weinheim, 1963, p. 1210.
[10] Friel, P. J., and Goetz, R. C. *J. Phys. Chem.*, 64 (1960) 175.
[11] Benson, S. W. *J. Chem. Educ.*, 42 (1965) 502.
[12] Spinks, J. W. T., and Woods, R. J. *Introduction to Radiation Chemistry*, 3rd ed. Wiley, New York, 1990.
[13] Freeman, G. R. In Freeman, G. R. (ed.), *Kinetics of Nonhomogeneous Processes*. Wiley, New York, 1987, p. 19.
[14] Paretzke, H. G. In Freeman, G. R. (ed.), *Kinetics of Nonhomogeneous Processes*. Wiley, New York, 1987, p. 89.
[15] Buxton, G. V. In Farhataziz and Rodgers, M. A. J. (eds.), *Radiation Chemistry: Principles and Applications*. Verlag Chemie, Weinheim, 1987, p. 321.
[16] Magee, J. L., and Chatterjee, A. In Freeman, G. R. (ed.), *Kinetics of Nonhomogeneous Processes*. Wiley, New York, 1987, p. 171.
[17] Henglein, A., Schnabel, W., and Wendenburg, J. *Einführung in die Strahlenchemie*. Verlag Chemie, Weinheim, 1969.
[18] LaVerne, J. A., and Pimblott, S. M. *J. Chem. Soc. Faraday Trans.*, 89 (1993) 3527.
[19] De Visscher, A., van Eencoo, P., Drijvers, D., and Van Langenhove, H. *J. Phys. Chem.*, 100 (1997) 11636.
[20] Hickling, R., and Plesset, M. S. *Phys. Fluids*, 7 (1964) 7.
[21] Fujikawa, S., and Akamatsu, T. *J. Fluid Mech.*, 97 (1980) 481.
[22] Apfel, R. E. *Methods Exp. Phys.*, 19 (1980) 355.
[23] Suslick, K. S., Hammerton, D. A., and Cline, R. E. J. *J. Am. Chem. Soc.*, 108 (1986) 5641.
[24] Shima, A., Tomita, Y., and Ohno, T. *J. Fluid Eng.*, 110 (1988) 194.
[25] Atchley, A. A., and Crum, L. A. In Suslick, K. S. (ed.), *Ultrasound: Its Chemical, Physical, and Biological Effects*. VCH, Weinheim, 1988, p. 1.
[26] Suslick, K. S. In Suslick, K. S. (ed.), *Ultrasound: Its Chemical, Physical, and Biological Effects*. VCH, Weinheim, 1988, p. 123.
[27] Verrall, R. E., and Sehgal, C. M. In Suslick, K. S. (ed.), *Ultrasound: Its Chemical, Physical, and Biological Effects*. VCH, Weinheim, 1988, p. 227.
[28] Shima, A., and Tomita, Y. In Tien, C. L., and Chawla, T. C. (eds.), *Annual Review of Numerical Fluid Mechanics Heat Transfer*. Hemisphere Publishing Corp. (Taylor and Francis), New York, 1989, p. 198.
[29] Young, F. R. *Cavitation*. McGraw–Hill, New York, 1989.
[30] Sochard, S., Wilhelm, A. M., and Delmas, H. *Ultrasonics Sonochem.*, 4 (1997) 77.
[31] Mason, T. J., and Lorimer, J. P. *Sonochemistry: Theory and Uses of Ultrasound in Chemistry*. Ellis Horwood, Chichester, 1988, p. 1.
[32] Jeffries, J. B., Copeland, R. A., Suslick, K. S., and Flint, E. B. *Science*, (1992) 248.
[33] Sochard, S., Wilhelm, A. M., and Delmas, H. *Ultrasonics Sonochem.*, 4 (1997) 77.
[33a] Colussi, A. J., Weavers, L. W., and Hoffmann, M. R. *J. Phys. Chem.*, 102 (1998) 6927.
[34] Lauterborn, W., and Ohl, C.-D. *Ultrasonics Sonochem.*, 4 (1997) 65.

[35]   von Sonntag, C. *The Chemical Basis of Radiation Biology*. Taylor and Francis, London, 1987.
[36]   Fricke, H., and Hart, E. J. *J. Chem. Phys.*, 3 (1935) 60.
[37]   Schuler, R. H., Patterson, L. K., and Janata, E. *J. Phys. Chem.*, 84 (1980) 2088.
[38]   Buxton, G. V., and Stuart, C. R. *J. Chem. Soc. Faraday Trans.*, 91 (1995) 279.
[39]   Anbar, M., and Pecht, I. *J. Phys. Chem.*, 68 (1964) 1460.
[40]   Mark, G., Tauber, A., Laupert, R., Schuchmann, H.-P., Schulz, D., Mues, A., and von Sonntag, C. *Ultrasonics Sonochem.*, 5 (1998) 41.
[41]   Matula, T. J., and Roy, R. A. *Ultrasonics Sonochem.*, 4 (1997) 61.
[42]   Berlan, J., and Mason, T. J. *Adv. Sonochem.*, 4 (1996) 1.
[43]   Todd, J. H., *Ultrasonics*, 8 (1970) 234.
[44]   Jana, A. K., and Chatterjee, S. N. *Ultrasonics Sonochem.*, 2 (1995) 87.
[45]   Price, G. J., and Lenz, E. J. *Ultrasonics*, 31 (1993) 451.
[46]   Hart, E. J., and Henglein, A. *J. Phys. Chem.*, 91 (1987) 3654.
[47]   McLean, J. R., and Mortimer, A. J. *Ultrasound Med. Biol.*, 14 (1988) 59.
[48]   Mason, T. J., Lorimer, J. P., Bates, D. M., and Zhao, Y. *Ultrasonics Sonochem.*, 1 (1994) 91.
[49]   Gutierrez, M., and Henglein, A. *J. Phys. Chem.* 95 (1991) 6044.
[50]   Hart, E. J., and Henglein, A. *J. Phys. Chem.*, 89 (1985) 4342.
[51]   Entezari, M. H., and Kruus, P. *Ultrasonics Sonochem.*, 1 (1994) S75.
[52]   Entezari, M. H., and Kruus, P. *Ultrasonics Sonochem.*, 3 (1996) 19.
[53]   Weissler, A. *J. Acoust. Soc. Am.*, 25 (1953) 651.
[54]   Alippi, A., Cataldo, F., and Galbato, A. *Ultrasonics*, 30 (1992) 148.
[55]   Hua, I., and Hoffmann, M. R. *Environ. Sci. Technol.*, 30 (1996) 864.
[56]   Francony, A., and Petrier, C. *Ultrasonics Sonochem.*, 3 (1996) 77.
[57]   Hart, E. J., and Walsh, P. D. *Radiat. Res.*, 1 (1954) 342.
[58]   Matthews, R. W. *Radiat. Res.*, 83 (1980) 27.
[59]   Armstrong, W. A., Facey, R. A., Grant, D. W., and Humphreys, W. G. *Can. J. Chem.*, 41 (1963) 1575.
[60]   Fang, X., Mark, G., and von Sonntag, C. *Ultrasonics Sonochem.*, 3 (1996) 57.
[61]   Buxton, G. V., Langan, J. R., and Lindsay Smith, J. R. *J. Phys. Chem.*, 90 (1986) 6309.
[62]   Fang, X., Pan, X., Rahmann, A., Schuchmann, H.-P., and von Sonntag, C. *Chem. Eur. J.*, 1 (1995) 423.
[63]   Pan, X.-M., Schuchmann, M. N., and von Sonntag, C. *J. Chem. Soc. Perkin Trans. 2*, (1993) 289.
[64]   Merga, G., Schuchmann, H.-P., Rao, B. S. M., and von Sonntag, C. *J. Chem. Soc. Perkin Trans. 2*, (1996) 1097.
[65]   Allen, A. O., Hochanadel, C. J., Ghormley, J. A., and Davis, T. W. *J. Phys. Chem.*, 56 (1952) 575.
[66]   Buxton, G. V., Greenstock, C. L., Helman, W. P., and Ross, A. B. *J. Phys. Chem. Ref. Data*, 17 (1988) 513.
[67]   Bielski, B. H. J., Cabelli, D. E., Arudi, R. L., and Ross, A. B. *J. Phys. Chem. Ref. Data*, 14 (1985) 1041.
[68]   Dowideit, P. Ph.D. Thesis, Ruhr-Universität Bochum , 1996.
[69]   Blitz, J. *Fundamentals of Ultrasonics*. Butterworths, London, 1967, p. 1.
[70]   Petrier, C., Jeunet, A., Luche, J.-L., and Reverdy, G. *J. Am. Chem. Soc.*, 114 (1992) 3148.
[71]   Mues, A. In Dechema, Frankfurt/Main, *1st Circular "Arbeitsgruppe Sonochemie,"* 1994.
[72]   Watmough, D. J., Quan, K. M., and Shiran, M. B. *Ultrasonics*, 28 (1990) 142.
[73]   Henglein, A. *Naturwissenschaften*, 43 (1956) 277.
[74]   von Elbe, G., and Lewis, B. *J. Am. Chem. Soc.*, 55 (1933) 507.
[75]   Gmelins Handbuch der Anorganischen Chemie. *Sauerstoff*, 3rd ed. Verlag Chemie, Weinheim, 1958, p. 367.
[76]   Didenko, Y. T., Nastish, D. N., Pugach, S. P., Polovinka, Y. A., and Kvochka, V. I. *Ultrasonics*, 32 (1994) 71.
[77]   Suslick, K. S., Mdleleni, M. M., and Ries, J. T. *J. Am. Chem. Soc.*, 119 (1997) 9303.

[78] Henglein, A., and Kormann, C. *Int. J. Radiat. Biol.*, 48 (1985) 251.

[79] von Sonntag, C., Mark, G., Schuchmann, H.-P., von Sonntag, J., and Tauber, A. In Luche, J.-L., Balny, C., Bénéfice, S., Denis, J. M., and Pétrier, C. (eds.), *Book of Proceedings, Chemical Processes and Reactions under Extreme or Non-classical Conditions, Conference of the COST Chemistry Action D6, Chambéry Nov. 29–Dec. 1 1996.* European Commission, Luxemburg, p. 11, 1997.

[80] Hine, J., and Mookerjee, P. K. *J. Org. Chem.*, 40 (1975) 292.

[81] Landoldt-Börnstein. *Transportphänomene I (Viskosität und Diffusion).* Springer, Berlin, 1969.

[82] Riddick, J. A., Bunger, W. B., and Sakano, T. K. *Organic Solvents: Physical Properties and Methods of Purification.* Wiley, New York, 1986.

[83] Tauber, A. Ph.D. Thesis, Ruhr-Universität, Bochum, 1998.

[84] Tauber, A., Mark, G., Schuchmann, H.-P., and von Sonntag, C. *J. Chem. Soc. Perkin Trans. 2.* In press.

[85] Choudhury, T. K., Lin, M. C., Lin, C.-Y., and Sanders, W. A. *Combust. Sci. Technol.*, 71 (1990) 219.

[86] Schwarzenbach, R. P., Stierli, R., Folsom, B. R., and Zeyer, J. *Environ. Sci. Technol.*, 22 (1988) 83.

[87] Martino, C. J., and Savage, P. E. *Ind. Eng. Chem. Res.*, 36 (1997) 1385.

[88] Tauber, A., Schuchmann, H.-P., and von Sonntag, C. *Ultrasonics Sonochem.* In press.

[89] Tauber, A., d'Allesandro, N., Mark, G., Schuchmann, H.-P., and von Sonntag, C. In Tielum, A. and Neis, U. (eds.), *Ultrasound in Environmental Engineering.* TUHH Reports on Sanitary Engineering, Vol. 25, Hamburg Harburg, 1999, p. 123.

[90] Tauber, A., Schuchmann, H.-P., and von Sonntag, C. *Chem. Europ. J.* In press.

CAN SONICATION MODIFY THE REGIO- AND STEREOSELECTIVITIES OF ORGANIC REACTIONS?

Jean-Louis Luche and Pedro Cintas

OUTLINE

Advances in Sonochemistry
Volume 5, pages 147–174.
Copyright © 1999 by JAI Press Inc.
All rights of reproduction in any form reserved.
ISBN: 0-7623-0331-X

1. INTRODUCTION

Synthetic chemists and material scientists, even those not directly involved in ultrasound research, know the considerable potential of this nonconventional radiation. Faster reactions under mild conditions can be carried out with a low-cost equipment with only minimal adaptation and employing technical-grade solvents [1].

As if this were not enough, sonochemists have also discovered other effects. In particular, ultrasound tends to favor radical mechanisms at the expense of polar ones [2], leading to a change in reaction outcomes which has been termed *sono-chemical switching* [3], and the intriguing fact that ultrasound may modify the stereoselectivity of some organic reactions through mechanisms which are not yet known. Herein the term *stereoselectivity* refers to regio- and diastereoselective processes. Obviously, sonochemistry being based on symmetrical physical effects offers no means to induce "absolute" enantioselectivity. As we shall see, however, these important stereochemical considerations should be analyzed with caution (*vide infra*).

These sonostereoselections which remain largely unpredictable were never ex-amined before by sonochemists who were not even aware of their existence. The aim of this chapter is therefore to bring together a series of results in which ultrasound modifies the stereochemical outcome with respect to the silent process. These facts offer the opportunity to ascertain the influence of ultrasound on chemical reactions as well as to validate, invalidate, or complement some aspects of the current knowledge of cavitational theories. In the following sections we have classified the types of stereoselective ultrasonic reactions from a possible thermal or pressure origin. This approach is, however, somewhat speculative, and in many cases, the origin of the stereoselectivity is unclear or the expected stereochemical effect does not occur for unknown reasons.

Before these discussions, it seems advisable to present briefly a few basic notions on stereochemistry [4], which are not very familiar to some organic chemists, to most other chemists, and not at all to theoreticians and acousticians.

2. A FEW DEFINITIONS RELATED TO STATIC AND DYNAMIC STEREOCHEMISTRY

The simplest case of isomerism is that of *regio*isomerism since it can be described by using only planar structures. However, in many organic, inorganic, or biological molecules, the constituting atoms are not arranged in a plane, but in a three-dimen-sional space. This property has the consequence that with the same nature, number, and sequence of atoms, several structures can be formed, differing only by the spatial arrangement of the atoms, corresponding to the existence of *stereo*isomers (from a Greek root meaning solid).

2.1 Regioisomerism

The existence of isomers corresponding to different geometrical arrangements of groups of atoms in a plane can be illustrated by the examples shown in Scheme 1. Obviously these compounds have different structures, and hence different properties (solubilities, melting or boiling points, chemical stabilities).

Their behaviors with respect to physical phenomena (heating, pressurization, electrical fields) are necessarily different. From the sonochemistry viewpoint, they should react differently when submitted to cavitation. The reactions leading to their formation are competitive pathways, one of which predominates because of kinetic or thermodynamic factors (*vide infra*).

2.2 Static Stereochemistry

This is the simplest case of stereoisomerism, which is not always well understood, even by organic chemists (Scheme 2). It can be seen that these two compounds I(+) and I(−), termed *enantiomers*, with exactly the same composition are not identical since they cannot be superimposed on each other. As mirror images, they differ only by the spatial arrangement of the four necessarily different a, b, c, and d groups bound to the central carbon, that is, the "center of chirality."[1] I(+) and I(−) cannot be distinguished by symmetrical agents, i.e., symmetrical physical properties (e.g., melting or boiling points, molecular volume, IR, UV, or NMR spectra in achiral media), or symmetrical chemical reagents. As an illustration, let us consider our hands (Greek *cheir* hand) and a glove. If this glove is symmetrical (e.g., an ambidextrous medical glove), it can be worn indifferently by both hands. In contrast, an unsymmetrical glove (the unsymmetrical agent) is adapted only to one hand, evidencing their mutual dissymmetry. In other words, chirality can be revealed only by itself. The consequence for the sonochemist is that cavitation, which implies symmetrical physical

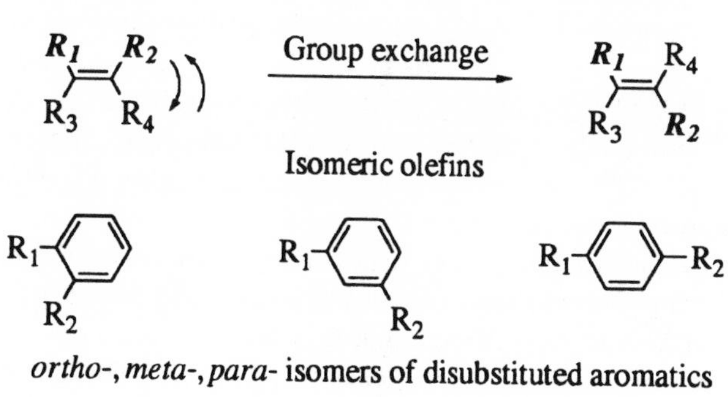

ortho-, *meta-*, *para-* isomers of disubstituted aromatics

Scheme 1. Olefinic and aromatic regioisomers.

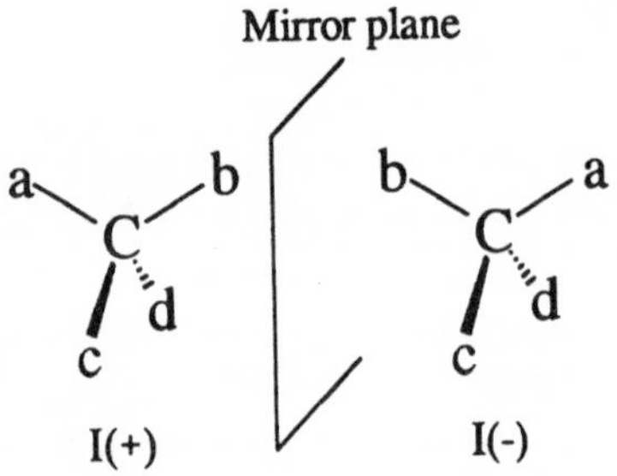

Mirror plane

Scheme 2. A pair of enantiomers.

phenomena, whether thermal, electrical, or concerned with pressure, is unable to distinguish between two enantiomers.

A racemic mixture consists of equal amounts of both enantiomers. A mixture of different amounts can be characterized by its enantiomeric excess (ee), calculated as the ratio of the difference between the concentrations of each species to their sum:

$$ee = [C_{I(+)} - C_{I(-)}]/[C_{I(+)} + C_{I(-)}]$$

When a second center of chirality is added, the situation is changed in an important manner according to the orientation of the groups around the second element (Scheme 3). II(+) and II(−) are still undistinguishable by symmetrical agents, being enantiomers [the same is true for III(+) and III(−)], but compounds II and III can no longer be deduced each from the other by a symmetry in a plane, i.e., they are not mirror images.

These compounds termed *diastereoisomers* (or *diastereomers*) have different properties, physical (molecular volume, melting or boiling point, electronic and nuclear spectra) and chemical (stability, reactivity). Obviously, the number of chirality elements in a molecule is not limited to two, and many natural products exhibit a large number of asymmetry centers; for proteins these can be thousands. An illustration among many others is that of the two sugars shown in Scheme 4, D-xylose (mp 156–158 °C, $[\alpha]_D$ 18.8) and D-lyxose (mp 108–112 °C, $[\alpha]_D$ 13.8). It can be predicted (and it would be of interest to obtain an experimental confirmation) that the physical effects of cavitation should have different consequences on these two diastereomers.

From a general viewpoint, the importance of stereochemical factors is primordial. The more striking consequence is certainly related to our health, and it is sufficient to observe that the biological properties of stereoisomers are very different in many cases. There are examples where one isomer is biologically (and beneficially) active while the second one is toxic. Most readers know the thalidomide story, which illustrates clearly the latter point. It is therefore of the highest importance for the

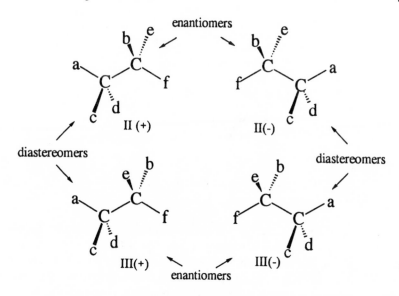

Scheme 3. The case of a molecule with two centers of chirality.

chemist to master selective methods permitting the preparation of the desired stereoisomer at will. This is the purpose of dynamic stereochemistry.

2.3 Dynamic Stereochemistry

The dynamic course of a reaction can be interpreted by the transition state theory, i.e., the pathway followed by reagents during a chemical transformation (Scheme 5).

From the energy level of the starting material(s), the system must acquire energy and go through a high point, the transition state. The difference in energy ΔE, the activation energy, has a direct influence on the reaction rate: the lower the value, the faster the process. When a reaction can potentially lead to two products, as is the case when two stereoisomers are formed, the system follows two parallel

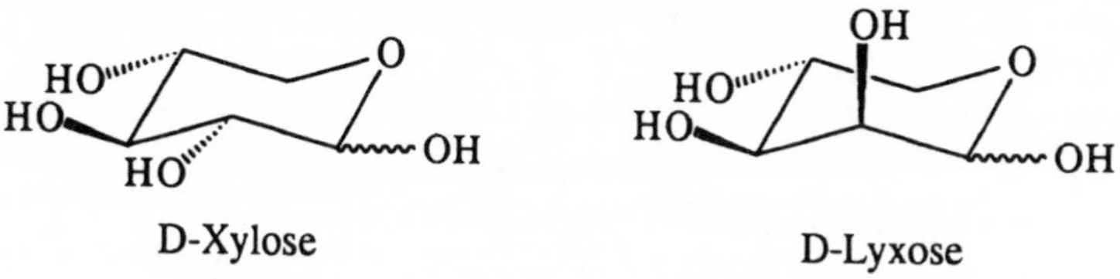

D-Xylose   D-Lyxose

Scheme 4. The diastereomeric pair of xylose and lyxose.

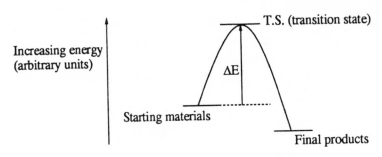

Scheme 5. The transition state interpretation.

(competitive) evolutions leading to two transition states, which have no reason (unless accidental) to reach the same energy levels (Scheme 6). The energy difference between these two transition states, $\Delta\Delta E$, leads to more or less stereose-lective reactions; the larger the difference, the higher the selectivity. A mixture of diastereoisomers is characterized by its diastereomeric excess (d.e.), which also measures the stereoselectivity of a reaction.

Here again, considering the sonochemist's viewpoint we will try to determine which factors are able to change the transition state energies.

With the help of Scheme 5, we see that the rate of a reaction can be increased either by increasing the initial energy level or by reducing the energy of the transition state. From an examination of the properties of cavitation, one can expect that the frequently observed sonication-induced rate enhancements can have both origins. Starting materials can be activated (i.e., enriched in energy) inside the cavitation bubble, thermally or by formation of any reactive intermediate (radical, radical ion). The second possibility, the lowering of the transition state energy, can be achieved e.g. by the generation of transient high pressures. Such high pressures

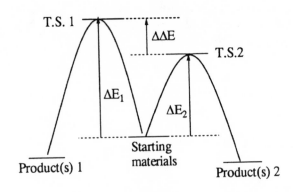

Scheme 6. The transition state interpretation of a reaction providing diastereomers.

bring the reagents into close proximity and thereby facilitate reaction. If stereoselectivity is considered, sonochemical alterations can occur if any of the effects of sonication is able to change $\Delta\Delta E$, the energy difference of the transition states.

The selective formation of stereoisomers (including regioisomers) can be described on these grounds when the reaction is under kinetic control. However, for reversible reactions where a thermodynamic control (i.e., the stability of the final products) modifies the product distribution, the important changes in the reaction kinetics induced by cavitation should result in a specific behavior for sonochemical reactions.

Even if there are relatively few papers in the literature reporting some sonostereochemical effect, they have two important consequences. The most significant is the discovery of synthetic methods with improved selectivities; the second consequence is indirect and consists of an expansion of our ideas on the relative importance of the physical phenomena associated with cavitation.

If from the synthetic viewpoint, common opinion holds that only the result is important, theoretical studies should be considered in order to establish unambiguously that such stereochemical effects are due to sonochemical causes. For this reason, in the reactions discussed below we will try to determine the role of sonication. This determination is not always easy, because sometimes there is only a very poor description of the experimental conditions employed. This is also a problem for sonochemists interested in the improvement of reproducibility.

3. ANALYSIS OF LITERATURE DATA

Probably the first case of a sonostereochemical effect found in the literature is the isomerization of maleic acid and its esters, by irradiation of their solutions in the presence of a source of halogen atoms (Scheme 7) [5]. Geometric stabilization of the radical adduct followed by elimination of halogen produces fumaric compounds.

This reaction was studied by several authors, and extended to various olefinic systems [6]. Mechanistically, the role of sonication is the formation of the actual reagent, the radical species, inside the bubbles, followed by the addition–isomeri-

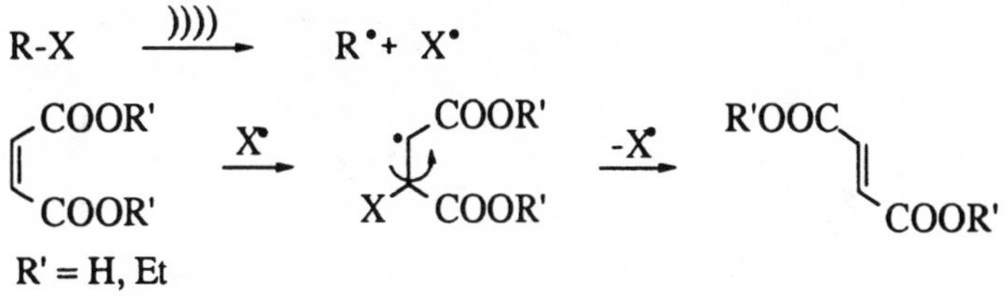

Scheme 7. Sonoisomerization of maleic compounds.

zation–elimination sequence which occurs in the bulk solution. Strictly speaking the overall process is not sonostereochemical in nature. Many cases discussed below correspond to a similar sequence: an initial sonochemical step occurring inside the bubble or in its vicinity, during which the sonochemical activation occurs, followed by a reaction in the solution wherein the stereoselection takes place, i.e., under conditions that can be different from the usual ones.

Since the origins of the sonochemical effect can be from various sources, we can draw a relatively unprecise line between the different examples, according to the supposed preeminence of the thermal, pressure, or other effects. This classification will hopefully be useful, though in the absence of further evidence it will have only an indicative value.

3.1 Effects Related to Temperature

Under conventional conditions, reaction rates and selectivities frequently respond in opposite manners to thermal activation. Slow processes are accelerated by heating, but the selectivities generally decrease. Knowing that the rate of sonochemical reactions can be increased by lowering the temperature [7], attempts were made to obtain simultaneously high rates, yields, and selectivities by sonication. Experimental results agree with this expectation in a number of reactions, for which the variation in the steric course can reasonably be assigned to the fact that the sonochemical reaction is advantageously conducted at a lower temperature.

3.1.1 Sonolytically Initiated Reactions

Japanese authors have described selective ultrasonic reactions of triaryltin hydrides [8]. The $R_3Sn\cdot$ radical formed in the bubbles adds to C–C multiple bonds in the bulk solution (Scheme 8), with sonostereoselectivities which may be assigned to the unusually low reaction temperatures. Thus, 1-hexyne gives the corresponding cis olefin in > 93%, which contrasts with the photolytic reaction affording a 1:1 cis/trans mixture.

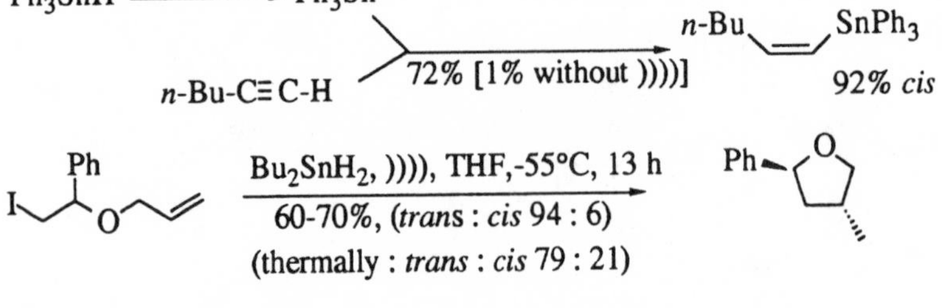

$$Ph_3SnH \xrightarrow{\text{))))},\ +7°C} Ph_3Sn\cdot$$

72% [1% without)))]

n-Bu-C≡C-H

n-Bu $=$ SnPh$_3$ 92% cis

Bu$_2$SnH$_2$,)))), THF, -55°C, 13 h

60-70%, (trans : cis 94 : 6)

(thermally : trans : cis 79 : 21)

Scheme 8. Stereoselection in tin hydride reactions.

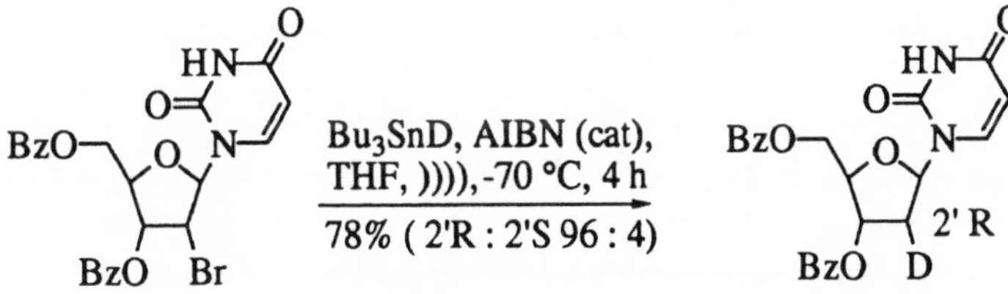

Scheme 9. Stereoselective deuteration of a C–Br bond.

The reduction of carbon–halogen bonds using di- and trialkyl (or triaryl) tin hydrides also proceeds at temperatures as low as –64 °C when sonicated. When unsaturation is present in the starting material, radical intermediates can be trapped to give cyclized product with improved stereoselectivities (Scheme 8). The advantages of this protocol were also evidenced in the reductive deuteration of bromo-ribonucleosides (Scheme 9) [9]. Refluxing the substrate and tributyltin deuteride in THF solution (ca. 65 °C) in the presence of AIBN, gives a mixture of epimers ($2'R$:$2'S$ = 82:18) in 93% overall yield. The sonochemical experiment at –70 °C provides the same products in a 96:4 ratio.

A second reaction in which the reactive species was expected to result directly from cavitation is the Pauson–Khand annulation which forms cyclopentenones in a single step from an alkyne, an alkene, and carbon monoxide with dicobalt octacarbonyl [$Co_2(CO)_8$] [10]. In analogy with the mechanisms of the sonochemical reactions of transition metal complexes [11], it can be accepted that the transition metal complex undergoes a metal–ligand bond cleavage in the bubble, then reacts out of it with the substrate.

This interpretation is, however, difficult to apply to the case shown in Scheme 10, since it would be surprising that such a substrate penetrates into the bubble in amounts sufficient to make the reaction proceed at a reasonable rate. One can just observe that the rate, yield, and selectivity are increased, in agreement with the formation of the active species in an energy-rich domain followed by a reaction in a "quiet zone" of the solution, but the exact origin of the effect is problematical.

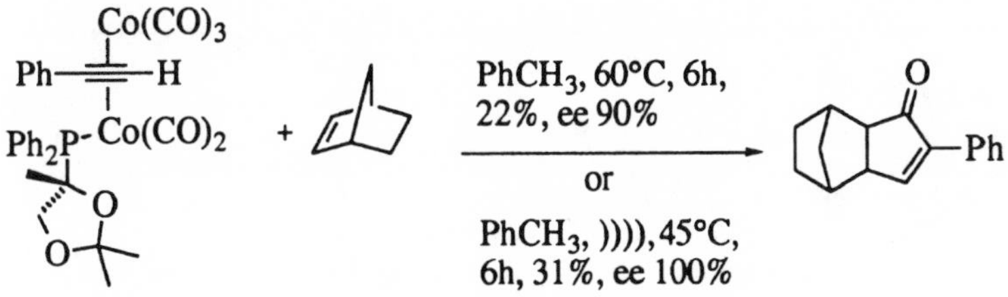

Scheme 10. Stereoselectivity of the Pauson–Khand annulation.

3.1.2 Reactions on Activated Surfaces

A series of reactions in which the reactive species need not be formed in the bubble exhibit some stereochemical effect. Examples were found for reactions forming C–C bonds and C–heteroatom bonds as well. The Barbier reaction using homochiral S(+)-2-halo-octanes constitutes an illustration (Scheme 11) [12]. The reactions of S-(+)-2-halo-octane proved to be strongly dependent on the nature of the halide, reflecting the relative rate of the reduction of the C–X bond.

The rate-determining electron transfer is sonication dependent. From the bromo compound, the reactive entity is the radical anion generated on the activated metal surface. Accelerating its formation, e.g., by using more efficient sonication conditions, increases its concentration and accelerates its addition to the carbonyl group, in a direction *anti* to the leaving bromide ion. As an element of comparison, the slow conventional reaction provides much lower yields of a practically racemic product.

A second illustration where ultrasonic energy is determinant is the reaction of sodium with 9-iodotabersonine, cyclized to the alkaloid vindolinine (Scheme 12) [13]. The reaction effected under high-energy sonication with a probe gives four stereoisomers,but only two from a less energetic irradiation.

The less energetic conditions (bath) produce only diastereoisomer 1, in a 30% yield. The result could at first glance be rationalized assuming that with low-energy irradiation the process occurs on the metal surface where the stereoselectivity results from the low degree of freedom of the adsorbed species. Desorption occurs when using a higher energy and the reaction in the bulk solution exhibits a diminished stereoselectivity. This interpretation agrees with the fact that ultrasound increases the mass transport at solid–liquid interfaces with an efficiency depending on the energy applied.

Among the reactions implying a nonmetallic surface, the preparation of γ-lactones from olefins involves the manganese triacetate oxidation of the

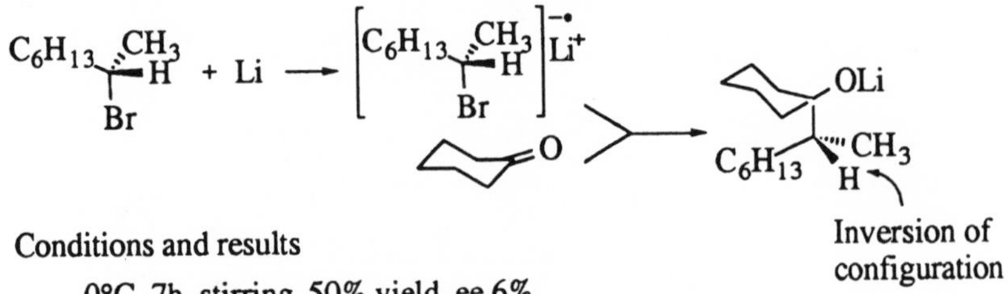

Conditions and results

    0°C, 7h, stirring, 50% yield, ee 6%
    -50°C, 1.5h,)))) medium energy, 98% yield, ee 19%
    -50°C, 1.5h,)))) high energy, 98% yield, ee 24%

Inversion of configuration

Scheme 11. Stereoselectivity of a Barbier reaction.

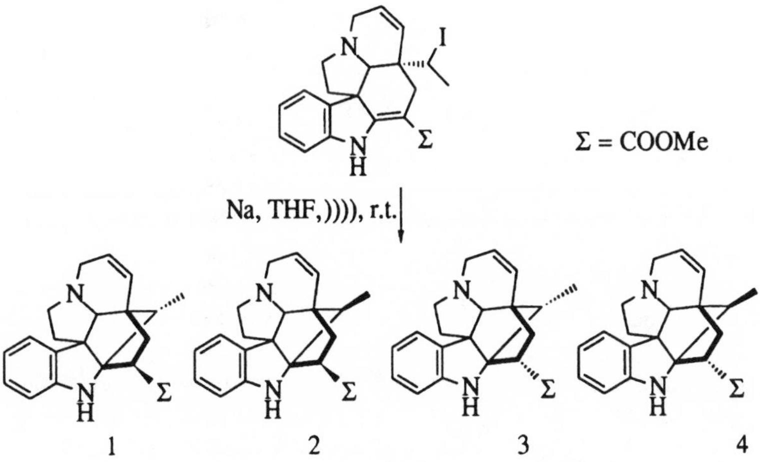

Scheme 12. Cyclization reaction of 9-iodotabersonine.

α-position of carbonyl groups, then the addition of the intermediate radicals to the olefinic bond. The reaction with monomethyl malonate in acetic acid, which does not occur at 0–10 °C, proceeds smoothly when sonication is applied [14]. From cyclohexene, only the *cis* ring fusion in the bicyclic lactone is observed (Scheme 13). The higher degree of stereoselectivity of the sonochemical process probably reflects the higher reaction rate, which does not allow equilibration processes to take place.

Asymmetric alkylations of a glycine equivalent derived from an Oppolzer's camphor sultam, leading to a practical preparation of enantiomerically pure α-amino acids, can be performed under PTC conditions with sonication (Scheme 14) [15]. The first experiments using vigorous stirring of glycine derivatives and allyl iodide in a system of dichloromethane–aqueous lithium hydroxide–tetrabutylam-

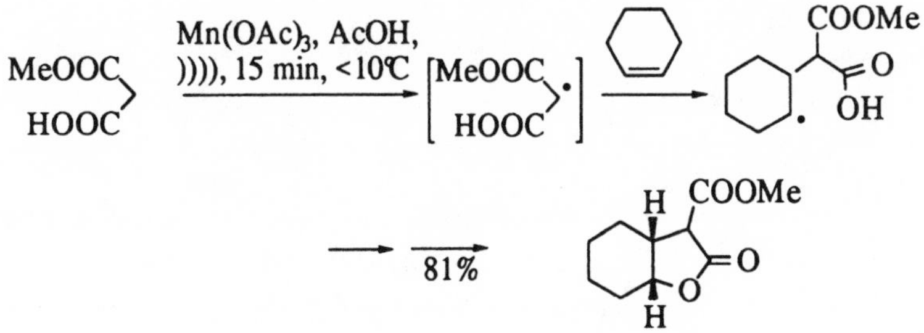

Scheme 13. Lactonization of cyclohexene.

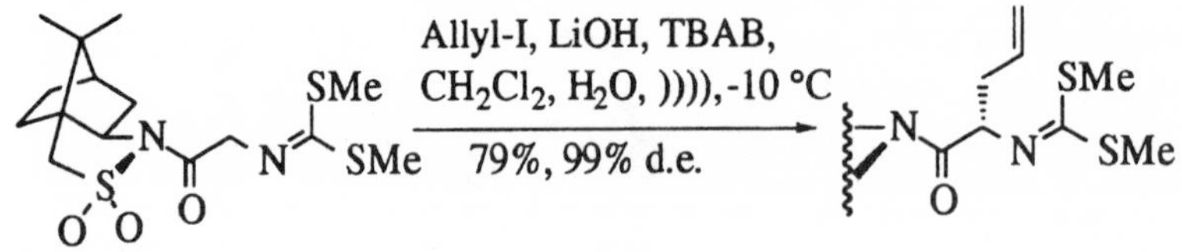

Scheme 14. Diastereoselective alkylation of a protected glycine element.

monium bromide, required 1–2 days for completion and suffered from competitive hydrolysis of the amide bond.

Sonication dramatically increases the alkylation rate, and a 79% yield of purified product was obtained after sonication for only 5 min at −10 °C. It can be envisaged that the acceleration of the rate-determining deprotonation, accompanied by an improved selectivity (reduced amide hydrolysis, higher diastereomeric excess) is explained by a microenvironment effect, as in the case of some [3+2] dipolar cycloadditions (*vide infra*).

Regiochemical effects can also be found in some instances, *inter alia* the addition of the trichloromethyl anion (from sodium trichloroacetate in acetonitrile) to quinoline and isoquinoline salts (Scheme 15) [16]. Conventionally, the reaction is accomplished by refluxing the mixture. The kinetically controlled reaction takes place at the 2-position, but the process is reversible and isomerization to the thermodynamic 4-trichloromethyl isomer occurs. In all of the cases, the rapidity of

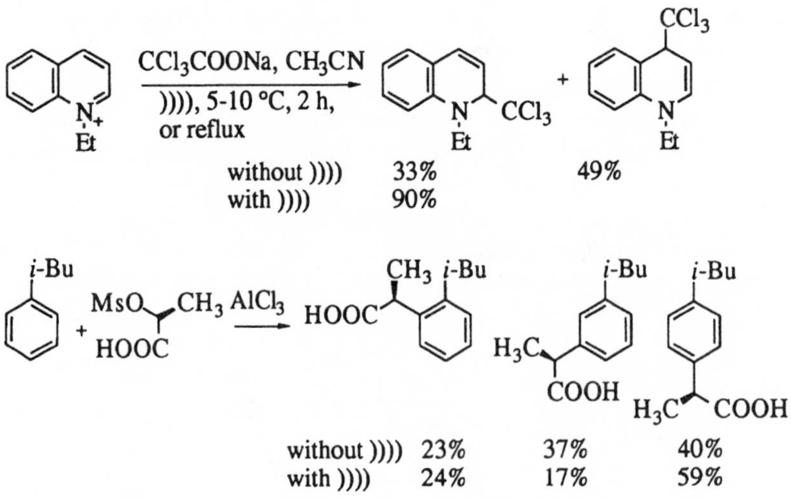

Scheme 15. Sonication-induced regioselectivity changes.

the sonochemical process achieved at a lower temperature means that the reaction will stop at the 1,2-addition stage despite its reversibility.

The Friedel–Crafts alkylation of isobutylbenzene was studied because of the interest in providing a nonsteroidal anti-inflammatory drug [17]. The change in regioselectivity induced by sonication is probably due to the same effect as in the previous example. It is known that the relative amounts of the kinetically formed *ortho*, *meta*, and *para* isomers can be modified by prolonged contact with the catalyst to reach concentrations in agreement with their thermodynamic stabilities [18].

Selectivities were observed in two related processes, the acetalization with long-chain aldehydes, ranging from hexanal to undecanal, and the glycosylation of unprotected sugars with long-chain alcohols (Scheme 16) [19, 20]. The glycosylation reaction is accelerated under sonication in the presence of heterogeneous catalysts such as silver or cadmium zeolites, or a mixture of silver perchlorate and carbonate on celite [21]. The stereospecificity depends on the nature of the alcohol acceptor, the nature of the support, and the solvent, but no major sonochemical improvement was evidenced. Silver zeolite provides the best results in 1,2-dichloroethane under sonication (α:β = 10:90, 80% yield). With unreactive alcohols, such as sugar alcohols, the strategy enables the preparation of disaccharides.

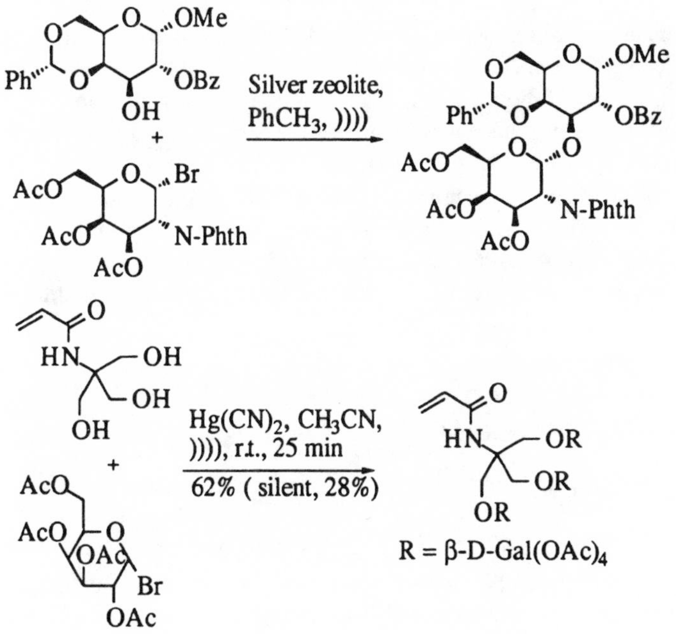

Scheme 16. Selective glycosylation reactions.

A glycosyl halide undergoes a clean substitution by an alcoholic group in the presence of mercury(II) cyanide. In comparison with the traditional method, the reaction is much faster and stereoselective [22]. Only the easily purified β-anomers are obtained, while the silent method gives mixtures.

3.2 Effects Related to Pressure

Interpretation of the pressure effects should be based on the preference of the reaction for the transition state with the smaller activation volume. Cycloadditions, many of which exhibit a strongly negative activation volume,[2] probably represent a paradigm case where pressure effects are expected to increase both the reaction rate and the selectivity, and in fact the application of high pressures has been of enormous benefit in cycloaddition chemistry [23]. Accordingly, the increase in pressure, and in temperature as well, provided by the cavitational collapse should have a dramatic effect on cycloadditions. Unfortunately, the experiment proves again that this is not always observed.

[2+2]Intramolecular reactions were reported to occur with improved yields and enantioselectivities when run under sonication at ambient temperature, in comparison with conventional procedures in refluxing solvents [24]. The reaction shown in Scheme 17 gives bicycloheptanone in 54% yield and 61% ee when sonicated. Conventionally the corresponding figures are 60% yield and 40% ee.

Such a change in stereoselectivity means that sonication is able to modify in some manner the energy difference between the diastereomeric transition states, but no reasonable explanation can be proposed.

In biphasic liquid–liquid [3+2] cycloadditions, practically no sonochemical influence on the steric course was found in most cases, in contrast to the sometimes important rate increases (Scheme 18) [25].

A recent study of the conventional [3+2] addition of nitrones to olefins demonstrated that changing the solvent from benzene to water substantially increases the rate without any effect on the stereoselectivity [26]. It can then be assumed that the role of sonication in the example of Scheme 18 is probably limited to that of an

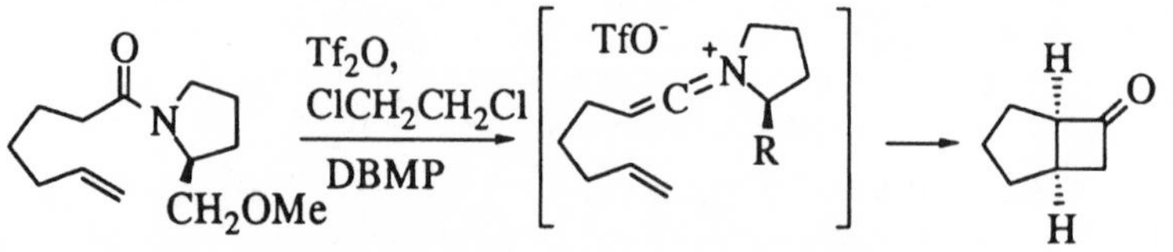

Tf : CF_3SO_2 ; DBMP : 2,6 di-t-butyl-4-methylpyridine
highest thermal ee : 40% (60% yield)
highest sonochemical ee : 61% (54% yield)

Scheme 17. Intramolecular cycloaddition of a keteniminium salt.

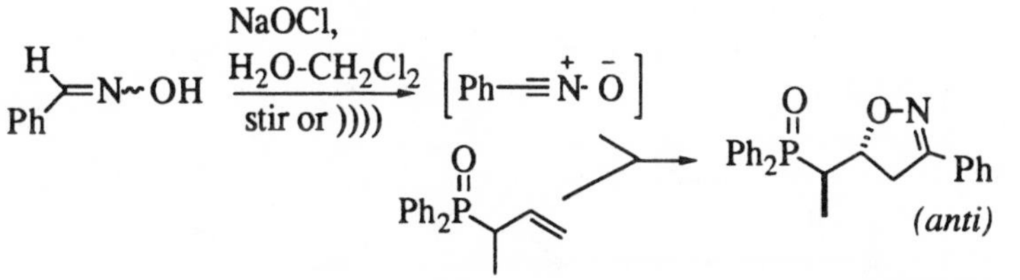

)))), r.t.,11 h, 78% (+ 15% *syn* isomer)
stir, r.t., 10 days, 72% (+ 21% *syn* isomer)

Scheme 18. Biphasic 1,3-dipolar cycloadditions.

efficient mixing. The reaction medium can be transformed to a fine emulsion approximating an aqueous medium in which the rates but not the stereoselectivity are affected.

In contrast, a similar cycloaddition effected in toluene only shows an interesting change in its regioselectivity by application of sonication [27]. The major difference with the previous case is the method of generation of the nitrile oxide, dehydro-chlorination of a chloro-oxime, and the medium, a purely organic solvent, but a mechanistic study would be necessary to determine if such parameters have a decisive importance.

Diels–Alder reactions are considered to proceed by concerted mechanisms and therefore constitute a lively scenario where ultrasound may eventually drive them through alternative mechanistic pathways. The existence of sonochemical effects with certain substrates (e.g., quinones) suggests that redox properties of reactants may be relevant [28], thereby facilitating the formation of radical ions, either from the diene or from the dienophile, as intermediates, even if the selectivity is not markedly affected. A systematic work exploring the experimental conditions of cycloadditions employing *ortho* quinones was reported (Scheme 19) [29].

The quinones are insoluble in an excess of diene and most of the time, the reactions were run under heterogeneous conditions. Under 11 kbar pressure or under sonication, the rates and yields are improved in comparison to thermal activation, and the isomeric ratio modified. The presence of a solvent, for instance toluene, makes the sonochemical effect disappear to a large extent. However, the presence of methanol in amounts just sufficient to homogenize the medium gives good results. That the solvent plays a direct role in the transition states is demon-strated by the change in regioselectivity when the polar methanol is replaced by the nonpolar dioxane [30]. A complete interpretation should take into account the actual activation volume changes of the solvated reaction partners.

A system related to the above consists of a *p*-benzoquinone and a styrene (Scheme 20). The thermal reactions in refluxing acetic acid give very low yields.

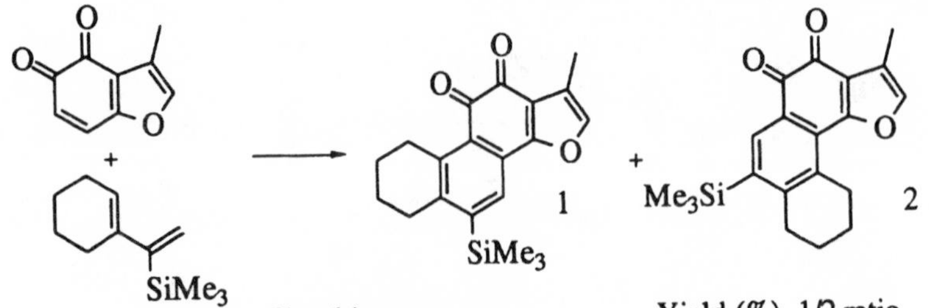

Conditions	Yield (%)	1/2 ratio
Ph, reflux, 12 h	18	3.5:1
MeOH, reflux, 4 h	28	3:2
MeOH, 11 kbars, r.t., 12 h	61	3.5:1
)))), neat, 45°C, 2h	57	5:1

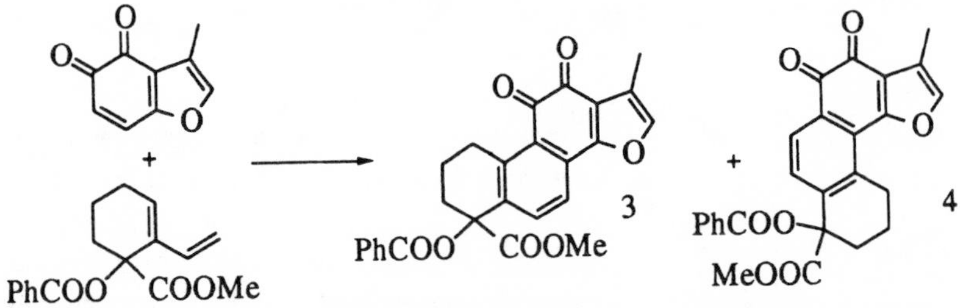

Conditions	Yield (%)	3/4 ratio
45°C,)))),neat,4h	56	1:1
id. without))))	45	1:1
45°C,)))), MeOH, 4 h	60	3.5:1
id. without))))	35	1.9:1
45°C,)))),dioxane, 4 h	37	1:2.6
id. without))))	10	1:1.2

Scheme 19. Yields and selectivity of a Diels–Alder reaction as a function of the conditions.

In contrast, sonication with zinc bromide in ethanol at room temperature leads to some success [31]. The yields remain modest at best, and some regioselectivity enhancement is recorded. We can observe here that the regioselection depends on a rather subtle discrimination between a methoxy and an allyl group, which are, furthermore, located rather far from the reaction site even in the favored *endo* transition state. Like the previous case the latter most probably includes not only the diene and dienophile but also the polar solvent and the Lewis acid catalyst.

Other examples of Diels–Alder cycloadditions where the kinetic effects of sonication are accompanied by stereochemical modifications have been published

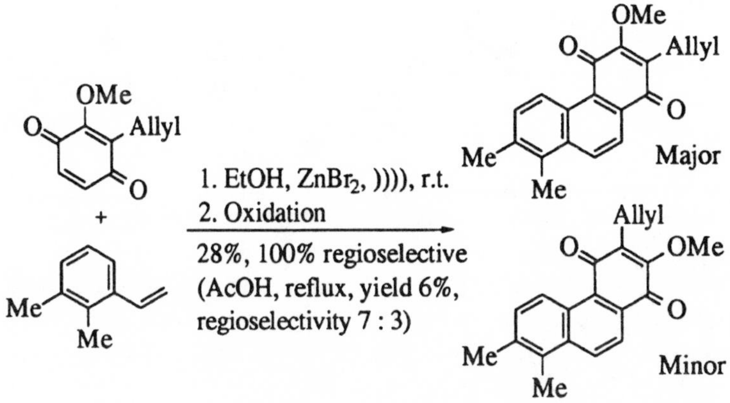

Scheme 20. Regioselectivity of a catalyzed Diels–Alder reaction.

[32, 33]. However, the experimental complexity (possible mechanistic changes, role of added Lewis acids and solvents), in addition to the sometimes poor descriptions make an interpretation, or even a classification, highly hazardous.

3.3 Effects of Undetermined Origin

In a number of cases where sonication modifies the steric course of a reaction, the interpretation is difficult: the temperature effects seem not to be involved directly, pressure effects are poorly known for these reactions, and kinetic data, which could explain the changes by a selective acceleration at some stage, are absent. Anyway, the sonostereoselectivity is interesting by itself when synthesis is considered.

3.3.1 Coupling Reactions

Deprotonation of an allylic phosphonium salt by butyl-lithium in a nonpolar solvent (Scheme 21) is very slow under stirring. Only 8% of the expected diene is obtained by addition of benzaldehyde after 2 h. The sonochemical reaction is not only much faster but also gives a quantitative yield, changing the *(Z)/(E)* ratio from 2:3 in THF to 1:3 in benzene [34].

The sonochemical effect may involve the formation of the phosphorus ylide in an unfavorable apolar medium where the insoluble lithium chloride is unlikely to be involved in the transition state. Changes in the nature of the reagent and/or the mechanism can be expected with probable effects on the stereoselectivity.

The improved control of the stereoselectivity in the cross-aldol reaction constitutes an important challenge [35]. Optimization was attempted by running the reaction sonochemically in the presence of alumina, without any solvent [36].

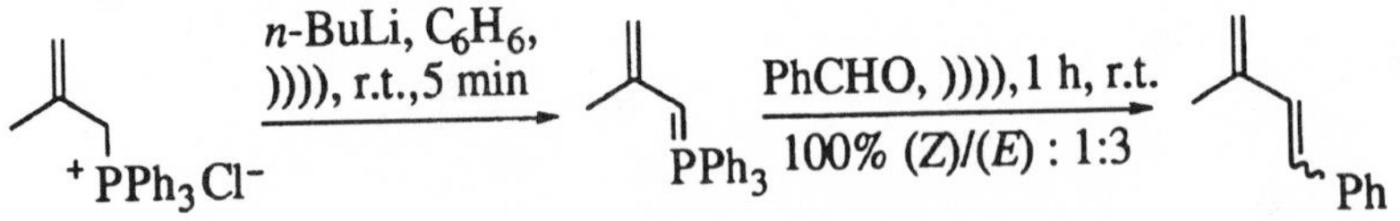

Scheme 21. Wittig condensation.

Sonication increases the rate and yields, but no direct mention is given of a stereochemical role. It is interesting to compare these results with those of the addition of phenoxides to glyceraldehydes, where the diastereomeric excess is increased from 60 to 92% by sonication (Scheme 22) [37].

A coupling catalyzed by manganese diiodide takes place from a nickelalactone and iodoalkanes (Scheme 23) [38]. The thermal reaction gives the coupled product in 44% yield after 24 h. Because of an equilibrium between the two isomeric structures of the organometallic, a 6:4 mixture of the linear-to-branched isomers is formed under silent conditions. Sonication leads to a 100% selectivity in favor of the linearly coupled product.

An explanation can involve a sonochemical equilibrium shift from the partial cleavage of the ligand–nickel bond, in analogy with known examples in transition metal chemistry [11]. The resulting low coordinated reagent should react preferentially at the ω-position.

Dianions of carboxylic acids can be prepared easily from the strong base lithium diisopropylamide (LDA) and the acid. When a leaving group, especially a chlorine atom, is present in the ω-position cyclization occurs in a few minutes at room temperature (Scheme 24) [39].

When applied to a dipeptide precursor, experiments using the sonochemically *in situ* generated LDA exhibit a higher stereoselectivity than with a conventionally prepared reagent. The main differences between the two methods are the presence

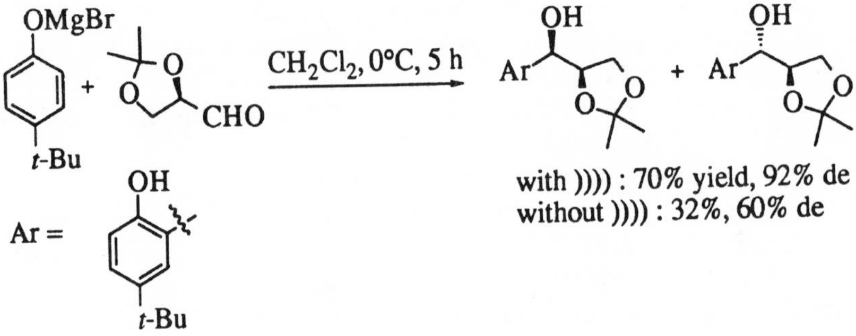

Scheme 22. Hydroxyarylation of a protected glyceraldehyde.

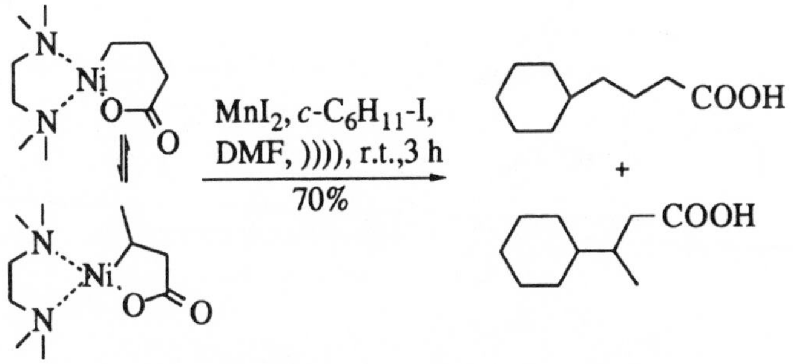

Scheme 23. Regioselective coupling of a nickelalactone.

of a metal and sonication in the sonochemical process. Which factor is responsible for the stereochemical result was not studied.

3.3.2 Reductions

Among the many organic reactions using metals, the reduction of carbonyl groups is frequently influenced by sonication. α-Diketones are reduced to enediol diethers in the presence of zinc and trimethylsilyl chloride. The rate increase is about tenfold, the yield is not improved, contrary to the stereoselectivity which is increased significantly (Scheme 25) [40].

In the presence of zinc in acetic acid–water solution, terpenic enone-lactones undergo a reduction of the carbon–lactonic oxygen bond (Scheme 26) [41]. When the latter is axial the conversion is completed in 2 min.

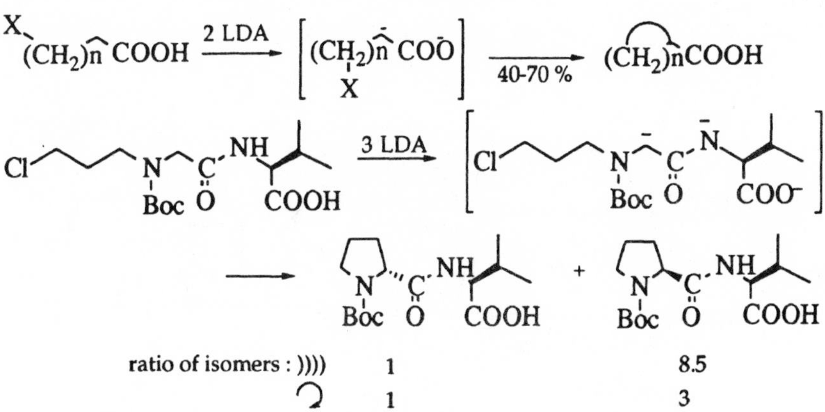

ratio of isomers :))))	1	8.5
$\cap$	1	3

Scheme 24. Stereoselectivity of cyclization of a dipeptide precursor.

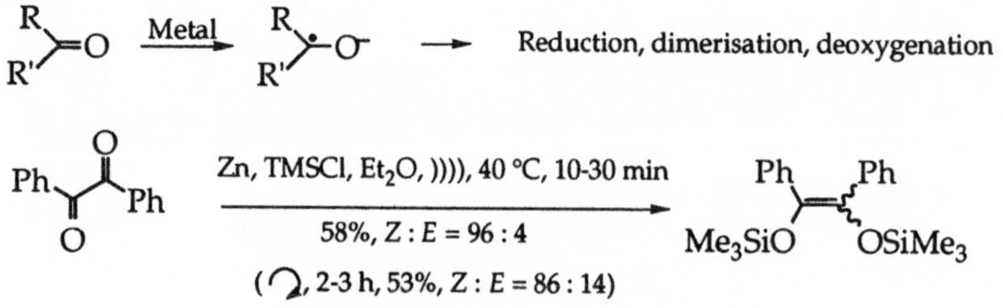

Scheme 25. Reduction of an α-diketone.

Conventionally the same reaction from the *trans* lactone occurs only after a preliminary isomerization to the *cis* compound. Under sonication, the cleavage requires longer irradiation times than for the *cis* lactone, but it occurs in good yields. This reaction will be interpreted only after its mechanism (sonoisomerization, sonoenolization, or reversible electron transfer to the conjugated system) is determined.

A few reactions are also relevant in this section, for instance the reduction of 2,5-dialkyl sulfolenes (Scheme 27) [42], or the cleavage of silicon–chlorine bonds [43] which leads to a coupling to new polymeric materials. The intermediate can be trapped by olefins to give silacyclopropanes stereoselectively. In each case a deeper understanding of sonochemical reaction mechanisms would help in the interpretation.

Similar conclusions can be extracted from the ultrasound-promoted lipase-catalyzed reactions [44]. Sonication enhances significantly reaction rates 7- to 83-fold, but stereoselectivity decreases under probe-ultrasonic conditions, perhaps by a temperature increase that favors racemization of the product or denaturation of the catalyst.

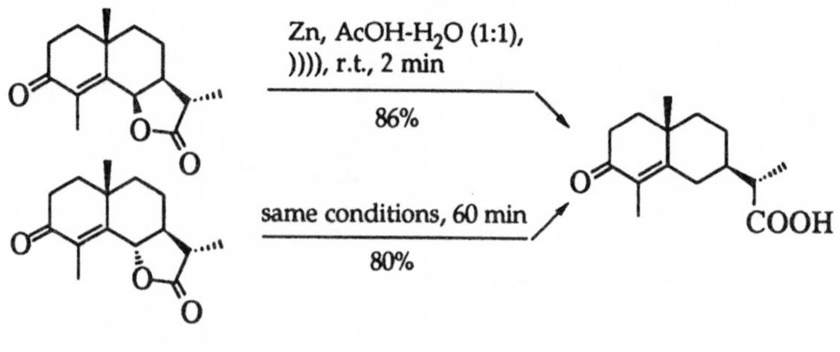

Scheme 26. Reductive ring opening of terpenic lactones.

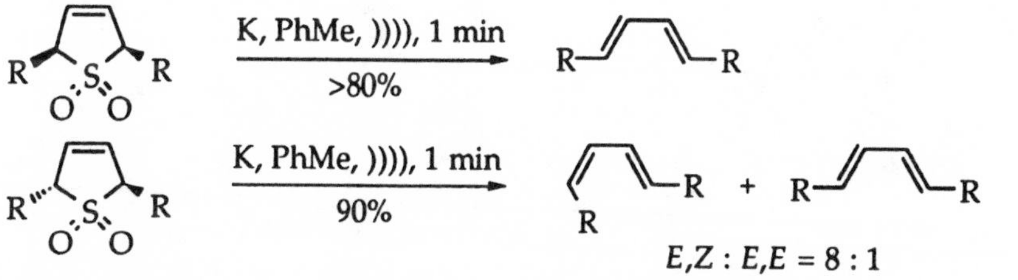

R = *n*-pentyl, *n*-hexyl, *n*-heptyl   With N_2 bubbling : $E,Z : E,E = 20 : 1$

Scheme 27. Ring cleavage of sulfolenes.

3.4 Reactions without Any Change

There are numerous examples in which the comparison of silent and sonochemical reactions suggests that the ultrasonic irradiation, sometimes unexpectedly, has little or no effect on the stereoselectivity. It would be useless and tedious to enumerate and discuss them, e.g., the examples given in Refs. 45–47. A few cases, however, exhibit an intrinsic interest which deserves some specific mention.

3.4.1 Reductions

In order to compare the stereoselectivity of the irradiated and nonirradiated reactions, the preliminary question is to ascertain if the same reaction intermediates are generated. Huffman et al. observed that the stereoselectivity of camphor reduction is the same either sonochemically in THF or silently in liquid ammonia with the three usual alkali metals (Scheme 28) [48].

The authors analyze the results to exclude a mechanism involving the ketone dianion. From a practical viewpoint, the sonochemical reductions in THF occur in 0.5–1 h, making this procedure much easier than the conventional reductions in liquid ammonia.

3.4.2 Addition of Organometallics

A number of examples involving fluorinated substrates have been published, and these have been reviewed [49]. With zinc as the metal, undesired side reactions that would occur with lithium or magnesium are avoided.

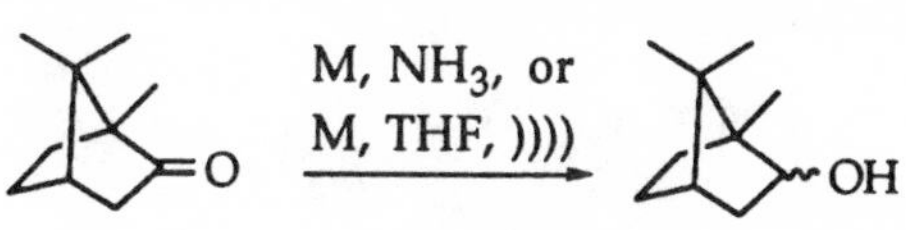

M	endo / exo ratio
K	4 / 6
Na	6.5 / 3.5
Li	7.5 / 2.5

Scheme 28. Alkali metal reductions of camphor.

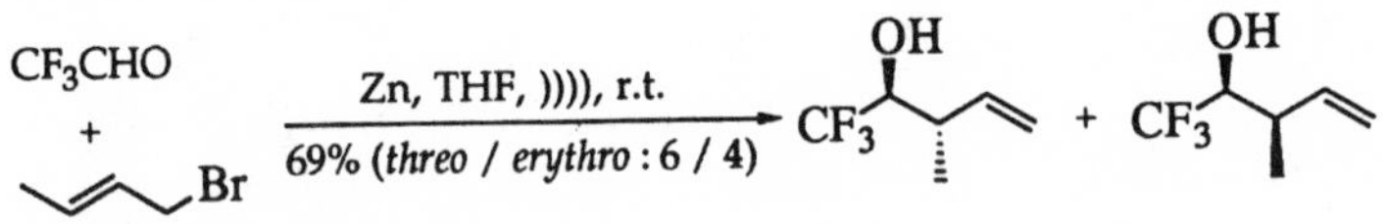

Scheme 29. Allylation of trifluoroacetaldehyde.

In the example shown in Scheme 29, the sonochemical reaction is easier and cleaner than the silent equivalent but the stereoselectivity is the same [50]. This is also the conclusion with respect to the Reformatsky reaction applied to imines [51, 52], and some of the "organometallic-like" reactions in aqueous media. Allylation of carbonyl compounds in a THF–aqueous ammonium chloride [53] system leads only to racemic compounds even when attempted in the presence of mannitol as a potential chiral inductor [54].

A similar absence of stereochemical difference between the Zn(Cu) couple and the thiohydroxamic ester pyrolysis method is mentioned in another work (Scheme 30) [55].

Explanations for the absence of any effect in these reactions are not easy to determine, except in a few cases. The disappointing results for the sonochemical addition of diethyl acetamidomalonate to chalcone (Scheme 31) in the presence of a chiral ephedrinium salt are caused by the limiting factor, the retro-Michael reaction. This process can more or less be enhanced by sonication, with a net zero balance for the effect on enantioselectivity [56]. We meet here a situation similar to those of the reactions described in Scheme 15.

That no ultrasonic influence can be observed when a strongly predominating factor is present is also illustrated in the Simmons–Smith cyclopropanation of allylic alcohols (Scheme 32) [57]. The stereoselectivity is similar to that of the

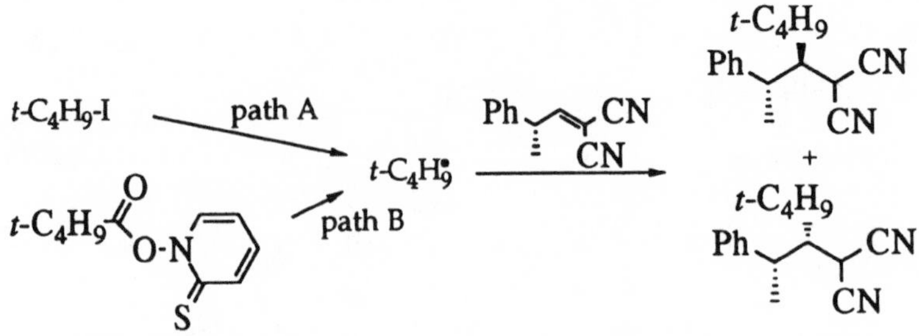

path A : Zn(Cu), H_2O,)))), 25 °C, 75%, *erythro* : *threo* 5.6 : 1
path B : $PhCH_3$, AIBN, 110 °C, 55%, *erythro* : *threo* 5.6 : 1

Scheme 30. Stereoselectivities of the radical conjugate additions to activated olefins.

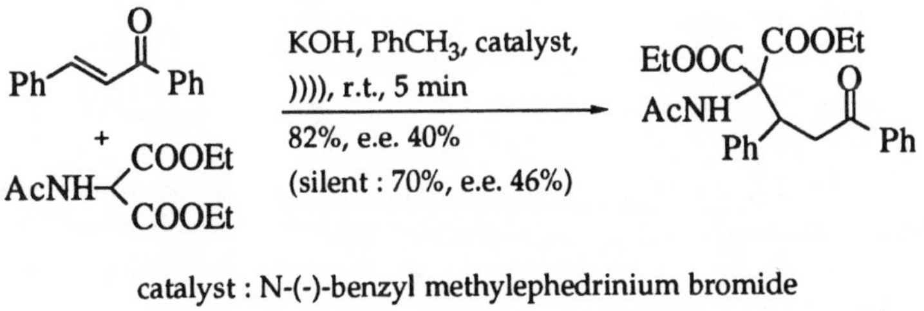

catalyst : N-(-)-benzyl methylephedrinium bromide

Scheme 31. Enantioselectivity of the addition of acetamidomalonate to chalcone.

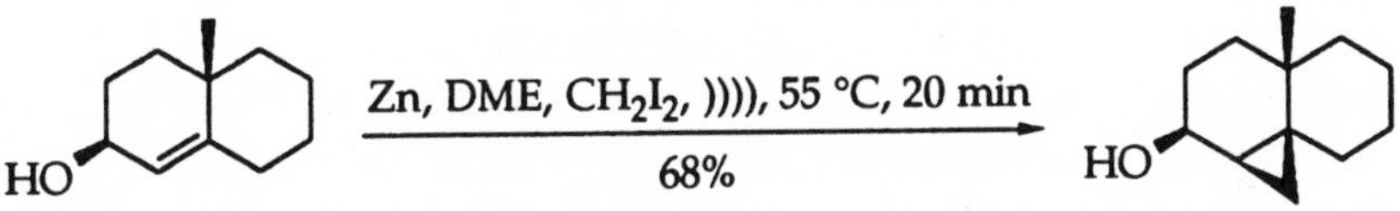

Scheme 32. Stereodirected Simmons–Smith cyclopropanation.

conventional reaction since the major stereodirecting influence exerted by the adjacent hydroxyl group is not overcome by any of the effects of sonication.

4. ELEMENTS OF DISCUSSION AND CONCLUSIONS

If a topic dominates synthetic organic chemistry by the end of this century, it must surely be the stereoselectivity of organic reactions. In recent years there have been considerable achievements in both enantio- and diastereoselective syntheses. The enantioselectivity can be reached by means of catalytic or absolute asymmetric syntheses in which chiral molecules are generated from achiral precursors. Absolute asymmetric synthesis has been accomplished in photochemical solid-state reactions and with the use of a chiral radiation. Diastereoselective syntheses can be improved by a variety of means. It is not surprising to see that "new" physical activation methods such as ultrasounds were used as physical agents able to lead to potential improvements. In the section below we will reexamine in more depth the possible origins of the sonostereochemical effect.

4.1 The Ultrasonic Field: A Falsely Chiral Influence with Autocatalysis?

Like other external energy sources, an acoustic field is not expected to directly induce chiral discrimination of racemates or absolute asymmetric synthesis.[3] This is a consequence of the "true chirality" condition, according to which neither uniform electric, magnetic, and gravitational fields nor nonpolarized electromagnetic radiation can induce molecular handedness when applied to isotropic systems that have been allowed to reach thermodynamic equilibrium [59]. However, for reactions under kinetic control, "falsely chiral" influences such as a static electrical field (a time-even polar vector) collinear with a static magnetic field (a time-odd axial vector) or, equivalently, a static magnetic field with oriented reactant molecules, might induce handedness in the products [60].

Although the local temperatures and pressures in the cavities behave as time-even scalars, the associated gradients ∇T and ∇P will be time-even polar vectors and could supplement any local electromagnetic fields as components of falsely chiral influences. So far, however, no enantioselectivity has been found using only an ultrasonic field as a falsely chiral source. Examples in which the ultrasonic irradiation improves the diastereoselectivity have been discussed above. A central question is the following: Does ultrasonic irradiation increase the already existing energy difference between the two diastereoisomeric transition states ($\Delta\Delta E$, Scheme 5) thereby further enhancing the preference to one of the isomers, or does it provide new mechanisms such as chiral autocatalysis?

Chiral autocatalytic processes, which occur under nonequilibrium conditions, are capable of amplifying small initial ee's to large final values [61]. In a recent and stimulating work, it has been theoretically demonstrated that a static magnetic field reduced the rate of a reaction step involving free radicals, through triplet state formation, thus allowing extra time for clusters of chiral catalyst–product molecules to reach a certain critical concentration [62]. Many ultrasonic reactions are radical processes, and a radical intermediate (not necessarily in its ground state) could function as described above. Moreover, since cavitation takes place far from equilibrium and can be accompanied by any type of electromagnetic component, together with large variations in temperature, pressure, and concentrations, there could be scope for the enhancement (or indeed the initiation) of chiral autocatalysis in sonochemical processes. Although this hypothesis has not yet been demonstrated, if sonostereoselective reactions, at least in some cases, are indeed chirally autocatalytic processes, this fact might account for both the rapidity and unexpected diastereoselection.

4.2 Conclusions

With these results it should be observed that when a sonostereochemical effect exists, it consists generally in an increase in the "conventional" stereoselectivity. A loss in selectivity can be observed less frequently, and no inversion was yet recorded. Considering the cavitation bubble as the sonochemical reactor [63], the determination of the reaction site is a key step for the knowledge of the sonochemical mechanism and vice versa. During collapse, a few nanoseconds, high temperatures (~5000 K) and pressures (~1500 bars) are produced in the cavity of bubbles. Although these effects constitute the basis of the "hot spot" theory [64], cavitation is apparently more than the simple addition of heat plus pressure. Some support exists in favor of alternative electrical theories claiming that intense electrical phenomena accompany the collapse [65, 66], and most of the sonochemical reactions cannot be interpreted in terms of only one hypothesis. The "hot spot" approach should then be better conceptualized under the form of a very high density of energy in a very small domain, even if the thermal form of energy is generally prevalent. This approach is valid for reactions occurring in the cavitation bubble. It is, however, important to point out that most of the reactions in which a stereo-chemical effect is observed imply nonvolatile substrates. The "hot spot" theory also states that considerable gradients of temperature and pressure exist on the limit layer of a collapsed bubble, from an internal temperature of several thousand degrees Kelvin to ambient temperature out of the bubble, this transition taking place over less than 500 Å. Substrates of low volatility cannot enter the bubble and so do not undergo directly the conditions of collapse in the bubble interior and no chemical, even stereochemical, effect should be observed. Then a possible explanation must take into account the pressure waves associated with the propagation of the acoustic wave or the shock waves generated during the bubble collapse. These pressure effects can propagate some distance from the bubble, and thus the reactions of even nonvolatile substrates can be affected. The sonostereochemical changes could then be the result of the role of pressure on the steric course of a reaction, the more compact transition state being favored. According to the general principles of organic reactivity, some processes will eventually be accelerated by very high pressures because of the large decrease in the volume of activation in forming the transition state. Sonication will also be useful when steric hindrance or the thermal instability of a reactant or product precludes the use of a conventional means to accelerate the reaction. Nevertheless, the latter considerations cannot be rigorously extrapolated to reactions sensitive to high pressures or temperatures when irradiated with ultrasound.

Even though hydrated electrons have not yet been unequivocally detected [67], the existence of charged species under these extreme conditions cannot be ruled out. Some scientists now believe that an electrical-like component, or more pre-cisely an electromagnetic one as a consequence of the motion of charged species, should also be present. What we have learned about the cavitational collapse and other hidden effects clearly reflects the complexity of this peculiar form of trans-mitting a high energy. Be that as it may, the study of the stereochemical changes in an acoustic field evidences not only the existence of external factors, but also that their combination has often a more profound effect on the energy of transition states than other conventional means of inducing reactivity. In this context it is noteworthy that diastereoselective reactions are influenced by a variety of factors including temperature, pressure, solvent, and concentration. However, if a unique source of radiation, providing these factors or others, whatever those may be, is capable of inducing such extreme stereoselectivities, this methodology will offer a substantial advantage over the often difficult combination of external factors.

In some cases the stereoselective enhancements may be attributed to a lower reaction temperature taking into account the anomalous temperature effect of ultrasonic reactions. Likewise, it appears evident that selectivity is a consequence of the thermal and pressure effects on the energy of transition states. While these observations account for the extreme conditions released during the cavitational collapse as well as the effects of other external factors on the cavitation itself, numerous cases cannot be interpreted with the current status of theory. Along with enormous temperature and pressure gradients from the microbubble to the bulk solution, another kind of electromagnetic component might also be involved. The overall effect would be the interaction of reactant molecules with a falsely chiral influence under kinetic conditions, and the propagation and enhancement of selectivity could also be due to an autocatalytic process when the favored isomer is formed at the expense of the minor one. This hypothesis, however, has not yet been elucidated.

ACKNOWLEDGMENTS

We are indebted to our co-workers whose names appear in the references. Special thanks go to Laurence D. Barron (Glasgow University) for his helpful comments on autocatalysis and stereochemical effects of external physical fields. Financial support from the Spanish D.G.I.C.Y.T. (PB95-0259-CO2-01) and the Junta de Extremadura-Fondo Social Europeo (PRI97-C175) is gratefully acknowledged.

NOTES

1. For the sake of simplicity, only the case of chirality centers is discussed. The more complex cases of axial or planar chiralities will not be presented here.
2. That is, the volume of the transition state is smaller than the total volume of the initial compounds.
3. It has been suggested that the resolution of racemates by preferential crystallization might be induced by ultrasonic irradiation in the range of 10–100 kHz [58].

REFERENCES

[1] For recent references in sonochemistry: Luche, J. L., and Cintas, P. In Fürstner, A. (ed.), *Active Metals: Preparation, Characterization, Applications*. VCH, Weinheim, 1996, pp. 133–190.
[2] Luche, J. L., Einhorn, C., Einhorn, J., and Sinisterra-Gago, J. V. *Tetrahedron Lett.*, 31 (1990) 4125–4128.
[3] Ando, T., and Kimura, T. In Mason, T. J. (ed.), *Adv. Sonochem.*, Vol. 2. JAI Press, London, 1991, pp. 211–251.
[4] Eliel, E. L., and Wilen, S. H. *Stereochemistry of Organic Compounds*. Wiley, New York, 1994.
[5] Elpiner, I. E., Sokolskaya, A. V., and Margulis, M. A. *Nature*, 208 (1965) 945–946.
[6] Caulier, T. P., Maeck, M., and Reisse, J. *J. Org. Chem.*, 60 (1995) 272–273. Peters, D., Pautet, F., El Fakih, H., Fillion, H., et al. *J. Prakt. Chem.*, 337 (1995) 363–367.
[7] De Souza-Barboza, J. C., Petrier, C., and Luche, J.-L. *J. Org. Chem.*, 53 (1988) 1212–1218.
[8] Nakamura, E., Imanishi, Y., and Machii, D. *J. Org. Chem.*, 59 (1994) 8178–8186.

[9] Kawashima, E., Aoyama, Y., Sekine, T., Nakamura, E., et al. *Tetrahedron Lett.*, 34 (1993) 1317–1320.

[10] Bladon, P., Pauson, P. L., Brunner, H., and Eder, R. *J. Organomet. Chem.*, 355 (1988) 449–454.

[11] Suslick, K. S., Goodale, J. W., Schubert, P. F., and Wang, H. H. *J. Am. Chem. Soc.*, 105 (1983) 5781–5787.

[12] De Souza-Barboza, J. C., Luche, J.-L., and Petrier, C. *Tetrahedron Lett.*, 28 (1987) 2013–2016.

[13] Hugel, G., Cartier, D., and Levy, J. *Tetrahedron Lett.*, 30 (1989) 4513–4516.

[14] Allegretti, M., D'Annibale, A., and Trogolo, C. *Tetrahedron*, 49 (1993) 10705–10714.

[15] Oppolzer, W., Moretti, R., and Zhou, C. *Helv. Chim. Acta*, 77 (1994) 2363–2380.

[16] Grignon-Dubois, M., Diaba, F., and Grellier-Marly, M. C. *Synthesis*, (1994) 800–804.

[17] Garot, C., Javed, T., Mason, T. J., Turner, J. L., et al. *Bull. Soc. Chim. Belg.*, 105 (1996) 755–757.

[18] March, J. *Advanced Organic Chemistry*, 4th ed. Wiley–Interscience, New York, 1992, p. 508.

[19] Miethchen, R., and Peters, D. *Z. Chem.*, 28 (1988) 298–299.

[20] Ferrieres, V., Bertho, J. N., and Plusquellec, D. *Tetrahedron Lett.*, 36 (1995) 2749–2752.

[21] Whitfield, D. M., Meah, M. Y., and Krepinsky, J. *Collect. Czech. Chem. Commun.*, 58 (1993) 159–172.

[22] Polidori, A., Pucci, B., Maurizis, J. C., and Pavia, A. A. *New J. Chem.*, 18 (1994) 839–848.

[23] For a survey of sonochemical cycloadditions: Fillion, H., and Luche, J.-L. In Luche, J.-L. (ed.), *Synthetic Organic Sonochemistry*. Plenum Press, New York, 1998, pp. 91–106.

[24] Chen, L., and Ghosez, L. *Tetrahedron Lett.*, 31 (1990) 4467–4470.

[25] Armstrong, S. K., Collington, E. W., Knight, J. G., Naylor, A., et al. *J. Chem. Soc. Perkin Trans. 1*, (1993) 1433–1447. Armstrong, S. K., Collington, E. W., and Warren, S. *J. Chem. Soc. Perkin Trans. 1*, (1994) 515–519.

[26] Pandey, P. S., and Pandey, I. K. *Tetrahedron Lett.*, 38 (1997) 7237–7240.

[27] Lu, T. J., and Sheu, L. J. *J. Chin. Chem. Soc.*, 42 (1995) 877–879.

[28] Nebois, P., Bouaziz, Z., Fillion, H., Moeini, L., et al. *Ultrasonics Sonochem.* 3 (1996) 7–13.

[29] Lee, J., and Snyder, J. K. *J. Org. Chem.*, 55 (1990) 4995–5008.

[30] Lee, J., Li, J.-H., Oya, S., and Snyder, J. K. *J. Org. Chem.*, 57 (1992) 5301–5312.

[31] Zhang, Z., Flachsmann, F., Moghaddam, F. M., and Ruëdi, P. *Tetrahedron Lett.*, 35 (1994) 2153–2156.

[32] Villacampa, M., Perez, J. M., Avendano, C., and Menendez, J. C. *Tetrahedron*, 50 (1994) 10047–10054.

[33] Carreño, M. C., Perez Gonzalez, M., and Fischer, J. *Tetrahedron Lett.*, 36 (1995) 4893–4896.

[34] Low, C. M. R. *Synlett*, (1991) 123–124.

[35] Gennari, C. In Trost, B. M. (ed.), *Comprehensive Organic Synthesis*, Vol. 2. Pergamon Press, Oxford, 1991, pp. 629–660.

[36] Ranu, B. C., and Chakraborty, R. *Tetrahedron*, 49 (1993) 5333–5338.

[37] Casiraghi, G., Cornia, M., and Rassu, G. *J. Org. Chem.*, 53 (1988) 4919–4922.

[38] Fischer, R., Walther, D., Bräunlich, G., and Undeutsch, B. *J. Organomet. Chem.*, 427 (1992) 395–407.

[39] De Nicola, A., Einhorn, C., Einhorn, J., and Luche, J.-L. *J. Chem. Soc. Chem. Commun.* (1994) 879–880.

[40] Boudjouk, P., and So, J. H. *Synth. Commun.*, 16 (1986) 775–778.

[41] Bargues, V., Blay, G., Cardona, L., Garcia, B., et al. *Tetrahedron Lett.*, 36 (1995) 8469–8472.

[42] Chou, T. S., and You, M. L. *J. Org. Chem.*, 52 (1987) 2224–2226.

[43] Boudjouk, P., Black, E., and Kumarathasan, R. *Organometallics*, 10 (1991) 2095–2096. Bahr, S. R., and Boudjouk, P. *J. Am. Chem. Soc.*, 115 (1993) 4514–4519.

[44] Lin, G., and Liu, H.-C. *Tetrahedron Lett.*, 36 (1995) 6067–6068.

[45] El Fakih, H., Pautet, F., Fillion, H., and Luche, J.-L. *Tetrahedron Lett.*, 33 (1992) 4909–4912.

[46] El Fakih, H., Pautet, F., Peters, D., Fillion, H., et al. *J. Prakt. Chem.*, 339 (1997) 176–178.

[47] Tomoda, S., and Usuki, Y. *Chem. Lett.*, (1989) 1235–1236.

[48] Huffman, J. W., Liao, W. P., and Wallace, R. H. *Tetrahedron Lett.*, 28 (1987) 3315–3318.

[49] Peters, D., and Miethchen, R. *J. Prakt. Chem.*, 337 (1995) 615–627.

[50] Kitazume, T. *Ultrasonics*, 28 (1990) 322–325.

[51] Bose, A. K., Gupta, K., and Manhas, M. S. *J. Chem. Soc. Chem. Commun.*, (1984) 86–87.

[52] Oguni, N., Tomago, T., and Nagata, N. *Chem. Express*, 1 (1986) 495–497.

[53] Cintas, P. *Synlett*, (1995) 1087–1096. Binder, W. H., Prenner, R. H., and Schmid, W. *Tetrahedron*, 50 (1994) 749–758.

[54] Einhorn, C., and Luche, J.-L. *J. Organomet. Chem.*, 322 (1987) 177–183.

[55] Roth, M., Damm, W., and Giese, B. *Tetrahedron Lett.*, 37 (1996) 351–354. Giese, B., Damm, W., Roth, M., and Zehnder, M. *Synlett*, (1992) 441–443. Urabe, H., Kobayashi, K., and Sato, F. *J. Chem. Soc. Chem. Commun.*, (1995) 1043–1044.

[56] Mirza-Aghayan, M., Etemad-Moghadam, G., Zaparucha, A., Berlan, J., et al. *Tetrahedron Asymmetry*, 6 (1995) 2643–2646.

[57] Clive, D. L. J., and Daigneault, S. *J. Org. Chem.*, 56 (1991) 3801–3814.

[58] Eliel, E. L., and Wilen, S. H. *Stereochemistry of Organic Compounds*. Wiley, New York, 1994, p. 311.

[59] Barron, L. D. *J. Am. Chem. Soc.*, 108 (1986) 5539–5542.

[60] Barron, L. D. In Cline, D. B. (ed.), *Physical Origin of Homochirality in Life*. American Institute of Physics, Woodbury, N.Y., 1996, pp. 162–182.

[61] Bolm, C., Bienewald, F., and Seger, A. *Angew. Chem. Int. Ed. Engl.*, 35 (1996) 1657–1659. Avalos, M., Babiano, R., Cintas, P., Jimenez, J. L., and Palacios, J. G. *Tetrahedron Asymmetry*, 8 (1997) 2997–3017.

[62] Hegstrom, R. A., and Kondepudi, D. K. *Chem. Phys. Lett.*, 253 (1996) 322–326.

[63] Leighton, T. G. *The Acoustic Bubble*. Academic Press, New York, 1993.

[64] Flint, E. B., and Suslick, K. S. *Science*, 253 (1991) 1397–1399. Henglein, A. In Mason, T. J. (ed.), *Adv. Sonochem.*, Vol. 3. JAI Press, London, 1993, pp. 17–83.

[65] Margulis, M. A. In Mason, T. J. (ed.), *Adv. Sonochem.*, Vol. 1. JAI Press, London, 1990, pp. 39–80.

[66] Lepoint, T., and Mullie-Lepoint, F. *Ultrasonics Sonochem.*, 1 (1994) S13–S22.

[67] Misik, V., and Riesz, P. *J. Phys. Chem.*, 101 (1997) 1441–1444.

THE USE OF ULTRASOUND IN MICROBIOLOGY

Sukhvinder S. Phull and Timothy J. Mason

Advances in Sonochemistry
Volume 5, pages 175–207.
Copyright © 1999 by JAI Press Inc.
All rights of reproduction in any form reserved.
ISBN: 0-7623-0331-X

1. INTRODUCTION

The use of ultrasound in microbiology and biotechnology is an expanding field but has its roots in one of the original uses of power ultrasound, namely, the disruption of biological cell walls to release the contents for *in vitro* studies. The type of apparatus used for this purpose is a horn resonating at 20 kHz dipped into a suspension of biological cellular material. This process depends on the efficiency with which the cell wall can be disrupted by the "ultrasound" to release its cellular contents without at the same time destroying them. This is much more difficult than it appears to be. The problem is that most simple one-cell organisms have an exceedingly tough cell wall only a few micrometers in diameter, and similar in density to the medium that surrounds it. The protein and nucleic acid components contained within the cell are large macromolecules, easily denatured by extreme conditions of temperature or oxidation both of which may arise from cavitation. A delicate balance must therefore be struck between the power of the probe and the disruption rate since power ultrasound, with its associated cavitational collapse energy and bulk heating effect, can denature the contents of the cell once released. Indeed for this type of usage it is important to keep the cell sample cool during sonication.

   One of the major areas of interest in the commercial application of sonochemistry in microbiology is a direct descendant of this type of work, namely, the destruction of microbiological material in water, i.e., disinfection without regard for the structural integrity of the contents. This is, however, only one of several important research interests regarding the possibility of using ultrasound in the biological sciences. As a result there are several research groups actively researching the use of ultrasound in the areas of microbiology and biotechnology.

   In this chapter we will concentrate on the use of ultrasound in the field of microbiology. The influence of the physical effects of ultrasound in microbiological processes will be described together with some recent applications of ultrasound to microbiology. For the benefit of readers who may not have an extensive knowledge of microbiology, the chapter begins by summarizing some of the types of organisms to be dealt with later.

2. SOME DEFINITIONS IN MICROBIOLOGY

Microbiology is the study of organisms that normally cannot be detected by the naked eye (microorganisms). These organisms include bacteria, viruses, certain algae, fungi, and protozoans. The existence of microorganisms was first demonstrated in the 1660s by Robert Hooke and Anton van Leeuwenhoek who developed microscopes and constructed powerful lenses, which were used to study and illustrate various microbes. The French chemist Louis Pasteur is considered the founder of microbiology. While Pasteur was not the first to argue that infectious diseases were caused by germs, his work was of paramount importance in demonstrating the relevance of germ theory to infectious disease, surgery, hospital management, agriculture, and industry [1].

2.1 The General Structure of Microorganisms

All microorganisms have some characteristics in common. The principal feature, and the one most directly affected by sonication, is the cell wall and/or cell membrane (see Figure 1). It is through the cell wall that all nutrient materials are absorbed and all excess or waste products are excreted. Some enzymes are produced and excreted in considerable quantities and this is a necessity for any organism that utilizes, for example, starch or cellulose. These molecules are large and must first be degraded by enzymes to a form in which they can be absorbed through the cell wall before they can be used by the cell. In some cases secondary metabolites are excreted in large amounts and in such cases they may be useful as antibiotics, e.g., penicillin, or harmful, as in the case of exotoxins which cause food poisoning.

Any process that affects the cell wall/membrane of a microorganism will clearly alter such processes. Ultrasound can improve mass transport and membrane permeability (see below) and both processes could improve microbial action. On the other hand, damage will cause an imbalance in the nutrients/toxins entering or leaving the microorganism, which in turn leads to cellular poisoning or starvation and cell wall rupture will kill the organism.

2.2 Types of Microorganisms

The definition of microorganisms as those organisms too small to be seen by the naked eye encompasses a large number of different types. With higher organisms it is convenient and easy to identify them as either plants or animals. Plants have rigid cell walls, are photosynthetic, and do not move independently. Animals have flexible cell walls, require organic food, and are capable of independent movement. This simple method of differentiation cannot be applied to microorganisms because of the simple structures of their cells and it has become convention to term all microorganisms *protists*. These can be divided into two types:

1. *Prokaryotes:* small (<5 μm) simple cell structures with rudimentary nucleus and one chromosome. Reproduction is normally by binary fission. Bacteria, actinomycetes, and the blue-green algae are included in this group.
2. *Eukaryotes:* larger (>20 μm) cells with a more complex structure and containing many chromosomes. Reproduction may be asexual or sexual and quite complex life cycles may be found. This class of microorganisms includes fungi, most algae, and the protozoa.

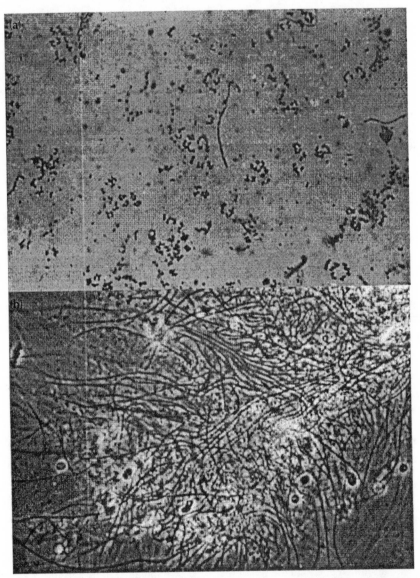

Figure 1. Photomicrographs of bacteria found in wastewater. (a) Typical bacteria dispersed in wastewater (×400). (b) Strands of *Sphaerotilus* with swarming cells (×400).

Table 1. Approximate Sizes of Microorganism Types

Organism	Size (μm)
Bacterium (rod)	(0.5–1.0) × (1.0–10)
Bacterium (sphere)	0.1–1.0
Fungus: yeast cell	(8–15) × (4–8)
Alga: *Chlamydomonas*	(28–32) × (8–12)
Virus: tobacco mosaic virus	0.3 × 0.015

There is a further group of microorganisms, the *viruses* which do not readily fit into either of the above classes and which are thus considered separately. The comparative sizes of the various microorganisms are given in Table 1.

2.2.1 Bacteria

Bacteria are simple, colorless, single-celled plants that use soluble food and may operate either as autotrophs or as heterotrophs and are capable of self-reproduction without sunlight. As species that are involved in the decomposition of matter, they fill an indispensable ecological role in the destruction of organic material in the natural environment and also, harnessed by man, in the handling of organic wastes in treatment plants. Bacteria range in size from approximately 0.5 to 5 μm and are, therefore, only visible through a microscope. Figure 1 shows photomicrographs of bacteria magnified 400 times. Individual cells may be spheres, rods, or spirals and may appear singly, in pairs, packets, or chains as shown in Figure 2. The basic structure and major components of a bacterium cell are shown in Figure 3.

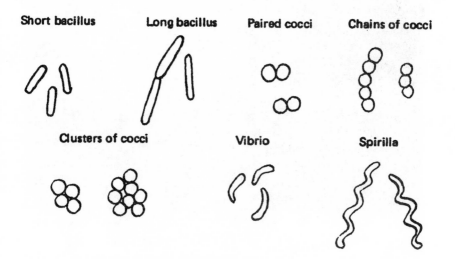

Figure 2. Common bacterial shapes.

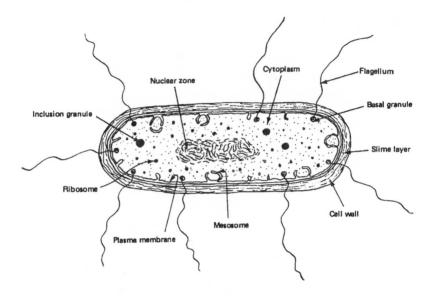

Figure 3. Basic structure of a bacterium.

Reproduction is by binary fission and the generation time for some species may be as short as 20 min under favorable conditions. Figure 4 shows the growth of the bacterium *Anthrobacter globiforms*. Some bacteria can form resistant spores that remain dormant for long periods in unsuitable environmental conditions but are reactivated on the return of suitable conditions (see next section).

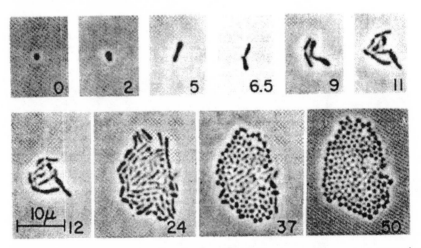

Figure 4. Photomicrographs of *Anthrobacter globiforms* growing on agar. Note the growth follows binary fission. The time of growth in hours is indicated on the photomicrographs.

Most bacteria prefer more or less neutral conditions of pH although some species can exist in a highly acidic environment. There are some 1500 known species, which are classified in relation to criteria such as size, shape, and grouping of cells; colony characteristics; staining behavior; growth requirements; motility, specific chemical reactions. Aerobic, anaerobic, and facultative forms are found. Bacteria are broadly classified into two major groups: heterotrophic and autotrophic, depending on their source of nutrients.

Heterotrophic bacteria (sometimes referred to as *saprophytes*) use organic matter as both a source of internal energy and sources of carbon for synthesis. These bacteria are further subdivided into three groups depending on their action toward free oxygen.

1. Aerobes require free dissolved oxygen in decomposing organic matter (referred to as *organics* below) to gain energy for growth and multiplication:

   Aerobic: organics + oxygen $\rightarrow CO_2 + H_2O$ + energy

2. Anaerobes oxidize organics in the complete absence of dissolved oxygen by using oxygen bound in other compounds, such as nitrate and sulfate.

   Anaerobic: organics + $NO_3^- \rightarrow CO_2 + N_2$ + energy

   organics + $SO_4^{2-} \rightarrow CO_2 + H_2S$ + energy

   organics $\rightarrow$ organic acids + $CO_2 + H_2O$ + energy

   organic acids + $CO_2 + H_2O$ + energy $\rightarrow CH_4 + CO_2$ + energy

3. Facultative bacteria comprise a group that uses free dissolved oxygen when available but can also live in its absence by gaining energy from anaerobic reaction. In waste treatment, aerobic microorganisms are found in activated sludge and trickling filters, but anaerobes predominate in sludge digestion. In essence, facultative bacteria are active in both aerobic and anaerobic conditions.

Heterotrophic bacteria decompose organics to gain energy for the synthesis of new cells, for respiration, and for motility. The amount of energy available from a given quantity of organic matter depends on the oxygen source used in metabolism. The greatest amount is available when dissolved oxygen is used in oxidation, and the least energy yield is derived from strict anaerobic metabolism [2].

Autotrophic bacteria oxidize inorganic compounds for energy and use carbon dioxide as a carbon source. There are three main sources of oxidizable inorganic materials:

1. Nitrifying bacteria oxidize nitrogen in ammonium groups to nitrate in a two-step reaction as follows:

$$NH_3 + oxygen–(Nitrosomonas) \rightarrow NO_2^- + energy$$

$$NO_2^- + oxygen–(Nitrobacter) \rightarrow NO_3^- + energy$$

2. Sulfur oxidation is typified by the reaction:

$$H_2S + oxygen \rightarrow H_2SO_4 + energy$$

The product from this reaction is sulfuric acid, which is a major source of corrosion in sewers. The wastewater in sewers often turns septic and this releases hydrogen sulfide gas via the reaction:

$$organics + SO_4^{2-} \rightarrow CO_2 + H_2S + energy$$

The hydrogen sulfide is absorbed in the condensation moisture on the side walls and crown of the pipe. Here sulfur bacteria, able to tolerate pH levels of less than 1.0, oxidize the weak acid H_2S to strong sulfuric acid using oxygen from air in the sewer. The sulfuric acid formed reacts with concrete, reducing its structural strength.

3. Iron bacteria are autotrophs that oxidize the soluble inorganic ferrous iron to less soluble ferric:

$$Fe^{2+} (ferrous) + oxygen \rightarrow Fe^{3+} (ferric) + energy$$

The mamentous bacteria *Leptothrix* and *Crenothrix* deposit oxidized iron, $Fe(OH)_3$, in their sheath, forming yellow or reddish-colored slimes. Iron bacteria thrive in water pipes where dissolved iron is available as an energy source and bicarbonates are available as a carbon source. With age the growths die and decompose releasing foul tastes and odors [3].

2.2.2 Endospores and Cysts

Some bacteria have the ability to prolong their existence by forming endospores resistant to high temperatures and toxic chemicals. This is characteristic of certain species of rod-shaped bacteria of the genera *Bacillus* and *Clostridium*. These (endo)spores are not reproductive units like the spores of fungi or higher plants since one cell produces only one endospore that germinates to produce only one new cell.

The bacterial endospore is a form of suspended animation, and can germinate when conditions for growth return to ideal (Figure 5). Some of these spores have a very long life and a well-known example of this is anthrax spores, which can survive for over 50 years in soil. This longevity is due to an extremely resistant coating of the spore, which may consist of several layers. In addition, the abnormally low

Dormant spore

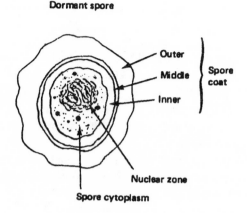

Outer

Middle ⎱ Spore
coat

Inner

Nuclear zone

Spore cytoplasm

Germinating spore

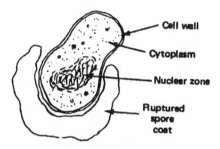

Cell wall

Cytoplasm

Nuclear zone

Ruptured
spore
coat

Figure 5. Structure of bacterial endospores.

water content of the spore cytoplasm (15% of vegetative cytoplasm) reduces the enzyme activity of the cell.

Some soil-dwelling bacteria such as *Azotobacter*, unable to form an endospore, round off and produce a cyst. The cell shortens and the cell wall becomes thickened. Cysts have limited powers of resistance to unfavorable conditions. On germinating they give rise to new vegetative cells.

2.2.3 Fungi

Fungi are aerobic multicellular plants that do not rely on photosynthesis for energy. They include yeasts and molds, which are more tolerant of both acid conditions and a drier environment than bacteria. They utilize much the same food sources as bacteria but their nitrogen requirement is less because their protein content is somewhat lower. Fungi form rather less cellular matter than bacteria from the same amount of food. They are capable of degrading highly complex organic compounds and some are pathogenic in man.

Yeasts are used industrially in fermentation in baking, wine making, brewing and from this the production of ethanol (by distillation). Fermentation processes, wherein yeast metabolizes sugar to produce alcohol, are carried out under anaerobic conditions because such conditions incur a minimum generation of new yeast cells. Under aerobic conditions, however, alcohol is not produced and the yield of new cells is much greater. For this reason the growth of yeast as animal feed using waste sugar or molasses employs aerobic fermentation.

Over 100,000 species of fungi exist and they usually have a complex structure formed of a branched mass of threadlike hyphae. They have four or five distinct life phases with reproduction by asexual spores or seeds. Fungi occur in polluted water and in biological treatment plants, particularly in conditions with high carbon-to-nitrogen ratios.

2.2.4 Actinomycetes

The actinomycetes are similar to fungi in appearance with a filamentous structure but with a cell size close to that of bacteria. They occur widely in soil and water and nearly all are aerobic. Their significance in water is mainly due to the taste and odor problems that often result from their presence.

2.2.5 Algae

Algae are plants, all of which use photosynthesis; most are multicellular although some types are unicellular. The majority of freshwater forms are the main producers of organic matter in an aquatic environment. Inorganic compounds such as carbon dioxide, ammonia, nitrate, and phosphate provide the food source to synthesize new algal cells and to produce oxygen.

Algae may be green, blue-green, brown, or yellow depending on the proportions of particular pigments within them. They occur as single cells that may be motile with the aid of flagella, nonmotile, or multicellular filamentous forms. Algae and bacteria growing in the same solution do not compete for food but have a symbiotic relationship in which the algae utilize the end products of bacterial decomposition of organic matter and produce oxygen to maintain an aerobic system. Algae of this type release toxins, which can be fatal to farm and domestic animals particularly if the animals drink water containing a significant amount of such algae. Toxins produced by blue-green algae can cause skin irritation to humans and may produce gastrointestinal illness; this problem has been particularly highlighted with the increasing popularity of water-contact recreation.

2.2.6 Viruses

Viruses are the simplest form of organism ranging in size from about 0.01 to 0.3 μm consisting essentially of nucleic acid and protein. They are all parasitic and cannot grow outside another living organism. All are highly specific regarding both the host organism and the disease that they produce. Human viral diseases include smallpox, infectious hepatitis, yellow fever, poliomyelitis, and a variety of gastrointestinal diseases. Identification and enumeration of viruses requires special apparatus and techniques.

2.2.7 Protozoa

Protozoa are unicellular organisms 1100 μm long that reproduce by binary fission. Most are aerobic heterotrophs and often utilize bacterial cells as their main food source. They cannot synthesize all of the necessary growth factors and rely on bacteria to provide these items. Protozoa are widespread in soil and water and may sometimes play an important role in biological waste-treatment processes. There are four main types of protozoa (Figure 6):

1. Sarcodina—ameboid flexible cell structure with movement by means of extruded pseudopod (false foot).
2. Mastigophora—motility by means of flagella.
3. Ciliatea—motility and food gathering by means of cilia (hairlike feelers). This type may be free swimming or attached to surfaces by stalks.

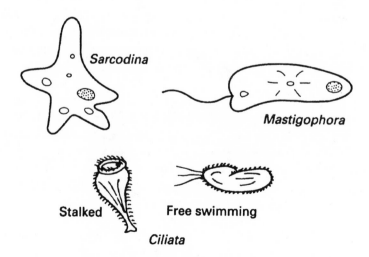

Figure 6. Typical examples of protozoa.

4. Sporozoa—nonmotile spore-forming parasites which are not found in water.

Most pathogenic protozoa, especially those found in water, are capable of forming spores or cysts (see earlier) that are highly resistant to common disinfectants and can thus be a source of water-borne infection even in developed countries with temperate climates.

3. THE EFFECTS OF ULTRASOUND ON MICROORGANISMS

Microorganisms can cause severe problems for human or animal health [4]. Disease-causing microorganisms, known as *pathogens*, can be present in the air, in water, and even in food. It is through water-borne microorganisms that a wide variety of diseases and illnesses are transmitted, some of which are listed in Table 2.

Table 2. Main Water-Related Diseases

Disease	Type of Water Relationship
Cholera	Waterborne
Infectious hepatitis	
Legionellosis	
Leptospirosis	
Paratyphoid	
Tularemia	
Typhoid	
Amebic dysentery	Water-washed or waterborne
Bacillary dysentery	
Gastroenteritis	
Ascariasis	Water-washed
Conjunctivitis	
Diarrheal diseases	
Leprosy	
Scabies	
Skin sepsis and ulcers	
Tinea	
Trachoma	
Guinea worm	Water related
Malaria	
Onchocerciasis	
Schistosomiasis	
Sleeping sickness	
Yellow fever	

It has been estimated that approximately 30,000 people die each day from water-related diseases and it should be pointed out that not all of these deaths occur in "developing" countries. Therefore, it is of the utmost importance to treat water efficiently and produce a final water that is both microbiologically and chemically safe for consumption as well as aesthetically acceptable. Scientists have been looking at applying ultrasound in conjunction with a number of other techniques as a means of destroying these pathogens and this topic will be described later, as will a range of ultrasonic disinfection/sterilization techniques.

Despite this "negative" image of microorganisms, it must be borne in mind that there are also a vast number of positive aspects to the action of these species the most basic of which is the recycling of vegetable matter by the decomposition of plants and animals with the subsequent return of nutrients to the soil or to the sea—a process on which all life depends. Over the years man has learned to make use of the synthetic and degradative abilities of microorganisms to provide useful chemicals. Fermentation processes are used not only in beer and wine production but also in the commercial synthesis of bulk chemicals such as citric acid (265,000 tonnes per annum) and L-glutamic acid (200,000 tonnes per annum). The total value of the world's production of synthetic microbial products is several millions of dollars [5].

3.1 Physical Effects of Sonomicrobiology

Before we look in detail at some of the applications of ultrasound in microbiology, it is important to understand how ultrasound interacts with microorganisms or even biological systems [6,7]. Regarding the possible application of ultrasound to biotechnology, Sinistera [8] describes in some detail, with specific examples, the possible effects and modes of action of ultrasound on biological systems. The cavitational bioeffects of continuous ultrasound at 1.5 MHz on three different classes of biosystems were investigated [9]. The retardation in the growth of broad bean (*Vicia faba*) plant roots, cell death, and DNA degradation in bacteria and human lymphocytes were reported [10]. A paper entitled "The Effect of Elastic Waves on the Life Activity of Some Prokaryote Bacteria" [11] reports a study into the effect of high- and low-frequency ultrasound on microbiological objects and defines informative variables pertaining to the activation or deactivation of the metabolism in prokaryote bacteria. It concludes that an ultrasonic field can affect the activity of DNA replication that occurs prior to cell division and that cavitation plays a very important role in this.

Ultrasound can affect microorganisms through a variety of mechanisms, some physical or mechanical and others more chemical in nature.

3.1.1 Temperature

Most ultrasonic experiments are carried out in temperature-controlled systems to ensure isothermal conditions so that bulk temperature effects will not influence any microbiological process [8]. A small general increase in microbial temperature can also influence both the active and passive transport systems of the cell membrane/wall [12,13] and this in turn may lead to an increased uptake of compounds. However, if the temperature is not controlled, then sonication could result in a large temperature increase (>60 °C). This in turn will lead to the denaturation (deactivation) of enzymes, proteins, and other cellular components present within the microorganism [14]. If sonication leads to cavitation within the cell, then this may cause the cytoplasmic proteins and/or DNA/RNA of microorganisms to become denatured and lead to mutation or death.

The heating effect of ultrasound has been utilized in a different situation for many years in physiotherapy, namely as an aid to massage for the treatment of various types of muscular strains. It can also be used to enhance the absorption of medicines through the skin although this is more related to the streaming effect of cavitation rather than heating. In one particular case, the uptake of hydrocortisone, applied externally to swine, increased by 300% using ultrasound [15]. Ultrasound is also capable of increasing the permeability coefficient of hydrocortisone in cellulose by 23% in an aqueous solution at 25 °C [16].

3.1.2 Cavitation

The mechanical effects of power ultrasound on chemical systems in a liquid medium are mainly attributed to cavitation and these same forces have a dramatic effect on biological systems. In general, cavitation, which in fact involves a complex series of phenomena, can be simplified into the following two types.

3.1.2.1. Transient cavitation. In transient cavitation, the cavitation bubbles, filled with gas or vapor, undergo irregular oscillations and finally implode. This produces high local temperatures and pressures, which would disintegrate biological cells and/or denature any enzymes present. The imploding bubble also produces high shear forces and liquid jets in the solvent, which may also have sufficient energy to physically damage the cell wall/membrane. Mechanical effects of this type have been used on a small scale for the disinfection of water contaminated with microbial spores, e.g., cryptosporidium, although the acoustic energy required is high [17,18].

3.1.2.2. Stable cavitation. This refers to bubbles that oscillate in a regular fashion for many acoustic cycles. The bubbles induce microstreaming in the surrounding liquid, which can also induce stress in any microbiological species present. This type of cavitation may well be important in a range of applications of ultrasound to biotechnology [19].

3.1.3 Standing Waves

In an opposite effect to the dispersion caused by power ultrasound referred to above, it is also possible to use ultrasound for particle agglomeration. Ultrasound, under the correct conditions, can produce standing waves in a liquid medium and small particles can be trapped within the antinodes of the wave [20]. The size of the particles that can be driven into these antinodes (and hence "separated") depends on the irradiation frequency $f(o)$ in hertz according to the simple formula:

$$f(o) = 0.48 \, n / R^2$$

where n is the liquid's kinematic viscosity (in $m^2 \, s^{-1}$) and R is the particle radius (in μm) (Table 3) [21]. When biological cells move into such bands, they aggregate into clumps [20,22]. These relatively large discrete clumps can then be removed from suspension in a controlled manner by modulation of the sound field [23–26].

3.1.4 Mass Transport and Mass Transfer through Cell Walls

An important consequence of the fluid microconvection induced by bubble collapse is a sharp increase in the mass transfer at liquid–solid interfaces. It has been demonstrated in the field of sonoelectrochemistry that ultrasound enhances mass transport to the electrode [27,28]. In microbiology there are two zones where this ultrasonic enhancement of mass transfer will be important. The first is at the membrane and/or cellular wall and the second is in the cytosol, i.e., the liquid present inside the cell.

3.1.4.1. Cellular membrane and wall. It has been demonstrated that ultrasound is able to enhance mass transfer through both artificial and biological membranes. Thus, the rate of NaCl transfer from 5% saline solution through a

Table 3. Relationship between Particle Size and Frequency for Particle Manipulation in an Acoustic Field

"Cut-off" Frequency	Diameter of Standing Particles (μm)
20 kHz	100
40 kHz	70
100 kHz	44
500 kHz	20
1 MHz	14
2 MHz	10
8 MHz	5
50 MHz	2
200 MHz	1
800 MHz	0.5

cellophane membrane into distilled water can be doubled when ultrasound is applied in the direction of diffusion [29]. The change in the rate of diffusion was not the result of either membrane damage or temperature variations in the medium. Similar effects have been reported for the diffusion of potassium oxalate and other solutes [15,16,30].

The effects of ultrasound on the permeability of the cell walls of the gram-negative bacterium *Pseudomonas aeruginosa* toward hydrophobic compounds, particularly antibiotics, have been examined [31]. The penetration and distribution of 16-dosylstearic acid (16-DS) in the cell membranes of the bacteria was quantified by a spin-labeling EPR method. The results indicated that the intracellular concentration of 16-DS was higher in insonated cells and increased linearly with the sonication power. EPR spectra revealed that ultrasound enhanced the penetration of 16-DS into the structurally stronger sites of the inner and outer cell membranes. The effect of ultrasound on the cell membranes was transient in that the initial membrane permeability was restored on termination of the ultrasound treatment. These results suggest that the resistance of gram-negative bacteria to the action of hydrophobic antibiotics was caused by a low permeability of the outer cell membranes and that this resistance may be reduced by the simultaneous application of antibiotic and ultrasound.

3.1.4.2. Cytosol. At an appropriate intensity level of ultrasound, intracellular microstreaming has been observed inside animal and plant cells with rotation of organelles and eddying motions in vacuoles of plant cells [32]. These effects can produce an increase in the metabolic functions of the cell, which could be of use in

$$H_2O \rightarrow OH\cdot + OH\cdot$$
$$OH\cdot + OH\cdot \rightarrow H_2O_2$$
$$OH\cdot + OH\cdot \rightarrow H_2O + O\cdot$$
$$OH\cdot + OH\cdot \rightarrow H_2 + O_2$$
$$H\cdot + O_2 \rightarrow HO_2^{\cdot}$$
$$HO_2^{\cdot} + H\cdot \rightarrow H_2O_2$$
$$HO_2^{\cdot} + HO_2^{\cdot} \rightarrow H_2O_2 + O_2$$
$$OH\cdot + H_2O \rightarrow H_2O_2 + O\cdot$$
$$H_2O\cdot + O\cdot \rightarrow H_2O_2$$
$$H\cdot + H\cdot \rightarrow H_2$$
$$H\cdot + OH\cdot \rightarrow H_2O$$

Scheme 1. Decomposition of water with power ultrasound.

both biotechnology and microbiology, especially in the areas of bioremediation and fermentation.

3.1.5 Free Radical Production

Cavitation induced in any liquid system will result in the formation of radicals [33]. In the case of water sonication gives rise to highly reactive radicals which can undergo a range of subsequent reactions as shown below in Scheme 1.

An important product from the sonolysis of water is hydrogen peroxide, which, together with the radical species, provides a powerful bactericide and chemical oxidant [34].

4. APPLICATIONS OF ULTRASOUND IN MEDICINAL MICROBIOLOGY

Ultrasound has been used as a diagnostic tool for many decades, and most people are aware of its use in fetal imaging. Additionally, ultrasound has become a useful diagnostic tool in many areas of medicine. As a direct result of this there are a number of research publications on the use of ultrasound in detecting severe and sometimes fatal diseases caused by microorganisms. For example, ultrasound can be used as a tool for the diagnosis of metastatic bacterial endophthalmitis which can lead to blindness [35], cervical and uterine infertility in the mare [36], renal lesions [37], and as an imaging tool to determine the nature of liver cysts caused by bacteria [38].

Other uses of ultrasound within the medical field have included the debridement of wounds contaminated with bacteria and/or particulate material [39]. The ability of current debridement techniques were compared with the relatively new ultrasound methodology to clean dorsal wounds on rats. In each case, 20 mg of montmorillonite clay, a well-known infection-inducing agent, was placed in the wound. The animals were randomly divided into three groups and the amount of clay removed from each wound using ultrasound debridement, soaking, and irrigation was measured. A further study combined a subinfective dose of *Staphylococcus aureus* bacteria and 10 mg of montmorillonite clay per wound and in these cases ultrasound debridement was compared with soaking, scrubbing, and high-pressure irrigation. The rats was examined after 7 days for inflammatory responses. Results of particulate contamination alone demonstrated that ultrasound debridement and irrigation remove statistically equal amounts of clay. In the second case, high-pressure irrigation and ultrasound debridement efficiently cleaned and disinfected contaminated wounds. Both series of experiments indicated that ultrasound debridement is an effective treatment of contaminated wounds and has been shown to be less traumatic.

Medical practitioners in California showed the effectiveness of ultrasound in the debridement of bone tissue [40]. Given the plethora of techniques available for debridement of contaminated bone, no single method can be considered ideal. The study was undertaken to evaluate sonication versus traditional debridement techniques in (1) their effectiveness in decontaminating trabeculated bone, (2) the subsequent effect of each treatment on bone cell function as measured by protein synthesis, and (3) the direct mechanical effects of each technique on the integrity of the bone structure itself. Ultrasonic debridement was found to be as effective as high-pressure jet irrigation or surgical scrubbing. Overall activated bone cell function 24 h after each debridement technique also was found to be equivalent. However, there was a radical difference in the effects of these treatments on exposed bone. Abrasive scrubbing and high-pressure jet irrigation both leave an exposed bone matrix not only devoid of any cells but also honeycombed with interstices for entrapment of bacteria and other contaminants. Only ultrasound maintained the integrity of the directly involved bone trabeculum to reduce contamination, prevent further colonization, and decrease further possible infection.

5. APPLICATIONS OF ULTRASOUND IN SEPARATION AND FILTRATION

Ultrasonic standing waves have the ability to move cells in suspension into bands in a standing wave field separated by a half-wave acoustic wavelength (see "Standing Waves" in Section 3). Bands and clumps can then be removed from suspension in a controlled manner by modulation of the sound field. In this way, yeasts have been removed efficiently from suspension in a system where the axis of the sample container was in the vertical plane and the standing wave nodal planes were horizontal [41].

The separation of erythrocytes (red blood cells) using a different configuration has been reported [42]. The suspension was held in a vertical container with the standing wave planes also vertical. The cells concentrated in these planes and formed clumps, which sedimented out to the bottom of the tube. The sedimentation rate of cell clumps formed in such vertical planes was found to be up to several thousand times faster than that of individual cells under gravity. This concept has been successfully exploited in the development of continuous flow "ultrasound-enhanced sedimentation" filters for the continuous filtration of hybridoma [43] and yeast [44].

Filtration efficiencies in excess of 99% have been achieved for the eukaryotic microbe *Saccharomyces cerevisiae* operating in a 10-MHz ultrasonic field [45]. The efficiency exceeded 95% for yeast concentrations between 1.5×10^7 and 4×10^9 mL^{-1}. The same group also studied the sedimentation rate of a suspension of bacteria and yeast cells in a continuous flow ultrasonic filtration system running at a frequency of either 1 or 3 MHz [46]. The filtration efficiency of *Escherichia coli* was greatest at 3 MHz (80%) whereas that of *S. cerevisiae* was greater than 99% at both frequencies. However, when mixed the filtration efficiency of *E. coli* fell to 50% due to particle interactions between the cells. Ultrasound has also been applied

Table 4. Pressure Amplitudes (PO) and Ramp Rates (n) Used in Transfer Tests (Downward or Upward Motion) on Microorganismsa

Organism	PO (kPa)	n (Hz) Down	n (Hz) Up
Saccharomyces cerevisiae	170	1	2
Bacillus megaterium	200	1	2
Listeria innocua	300	1	2–3
Lactococcus lactis	300	1	4–5
Escherichia coli	360	1	Not tested
Micrococcus luteus	370	1	2–3

Note: [a]Frequency of ultrasound 2.05 MHz with sawtooth frequency modulation of amplitude 33.3 kHz.

successfully to enhance the settling rate of filamentous bacteria in activated sludge [47].

Ultrasound in the frequency range 1–3 MHz has been used to move bacterial cells in pure culture and the findings suggest that it is possible to separate a range of bacteria of varying sizes and to concentrate them by ramping the applied ultrasonic frequency [26]. The paper demonstrated that ultrasound can be effectively used to band and concentrate very small bacteria (<0.5 μm diameter), such as *Micrococcus luteus* and *Listeria innocua* by ramping up or down the applied ultrasonic frequency. The authors used this technique to provide an effective separation of viable *E. coli* K12 in a 1% (v/v) milk solution. The methodology employed 2.05-MHz ultrasound at 360 kPa to establish first the banding and clumping in a stationary field. After about 3 min, the field was turned off to allow the white clumps of fat to float to the top of the cuvette. When this procedure was repeated, more fat material separated. After a further treatment time of 2 min, a negative frequency ramp was applied for 2–3 min at a ramp rate of 1 Hz, and gray bands of bacteria were formed and moved downward. This type of manipulation has been applied to a range of microorganisms with either upward or downward movements induced by frequency ramps (Table 4).

Recently a new method has been described for the selective isolation of species of *Myxococcus* directly from soil by dilution plating. Ultrasound treatment of soil suspensions gave the highest number of *Myxococcus* colonies in the soils studied [48].

6. MICROBIOLOGICAL DECONTAMINATION

High-power ultrasound has been known to disrupt biological structures for many years and in the early 1960s researchers began to formulate possible mechanisms for this effect (see Section 3) [49]. In 1975 it was shown that a brief exposure to ultrasound caused a thinning of the cell wall attributed to the freeing of the cytoplasmic membrane from the cell wall [50].

6.1 Decontamination of Aqueous Media

Several authors have shown that ultrasound has a disinfecting effect on microorganisms. The germicidal efficacy of ultrasound at a frequency of 26 kHz was evaluated by exposing aqueous suspensions of bacteria (*E. coli*, *S. aureus*, *Bacillus subtilis*, and *Pseudomonas aeruginosa*), fungus (*Trichophyton mentagrophytes*), and viruses (feline herpes virus type 1 and feline calici virus) to ultrasound [51]. There was a significant effect of time for all four bacteria, with percent killed increasing with increased duration of exposure, and a significant effect of increased kill with intensity for all bacteria (except *E. coli*). Positive results were also recorded for the reduction in fungal growth and reduction in feline herpesvirus with increased ultrasound intensity but there was no apparent effect on feline calicivirus. The results suggest that ultrasound in the low kilohertz frequency range is capable to some degree of inactivating certain disease agents that may reside in water.

From these studies and others it has been established that sonication can provide powerful disinfection, although 100% kill rates using only ultrasound require very high ultrasonic intensities [52]. Unfortunately, this makes the technique expensive to use for large-scale microbiological decontamination. However, over the last two decades, scientists have found that microorganisms are becoming resistant to the disinfection techniques involving chemicals, UV, and heat treatment and have once again started investigating ultrasound, but this time as an adjunct to other techniques.

Conventional methods of disinfection involve the use of a bactericide such as chlorine or ozone in the water industry. Despite the fact that chlorine has proved to be successful in combating a range of water-borne diseases, there are problems associated with its use:

- Microorganisms (especially bacteria) are capable of producing strains that are tolerant to normal chlorine treatment levels. This can be overcome by increasing the chlorine dose, although unpleasant flavors and odors can be generated by the formation of chlorophenols and other halocarbons through reaction with chemical contaminants in the water.
- Certain species of microorganisms produce colonies and spores that agglomerate in spherical or large clusters. Chlorination of such clusters may destroy microorganisms on the surface leaving the innermost organisms intact.
- Fine particles such as clays are normally removed by flocculation using chemicals such as aluminum sulfate. The flocs can entrap bacteria and spores and although the vast majority of floc particles are removed during processing it is possible that one or two may pass through the system and the bacteria protected by the floc material may well be unaffected by further disinfection.

Current trends are toward the reduction in use of chlorine as a disinfectant either by replacement with other biocides or by a reduction in the concentration required for treatment. Low-power ultrasound offers the latter possibility since it is capable of enhancing the effects of chemical biocides. The effect is thought to be due in part to the breakup and dispersion of bacterial clumps and flocs which renders the individual bacteria more susceptible to chemical attack. In addition, cavitation-induced damage to bacterial cell walls will allow easier penetration of the biocide.

The results of a study of the combined effect of low-power ultrasound and chlorination on the bacterial population of raw stream water are shown in Table 5. Neither chlorination alone nor sonication alone was able to completely destroy the bacteria present. It is significant that extending the time of chlorination or sonication from 5 min to 20 min seems to double the biocidal effect of the individual techniques. When sonication is combined with chlorination, however, the biocidal action is significantly improved. These results suggest that ultrasound could be used in conjunction with chemical treatments to achieve a reduction in the quantity of bactericide required for water treatment [34].

Zooplankton often accidentally pass through the purification cycle of a water treatment plant, leading to regermination and a clogging of filters located in the water distribution system. It is therefore important to eliminate the plankton before the water reaches the flocculation process. Such inactivation can be achieved using power ultrasound through the purely mechanical effects of acoustic cavitation as discussed above [53]. In order to inactivate plankton a sound intensity of approximately 1 W cm^{-2} and a high air content in the water are especially effective. The economic viability of a system for plankton treatment has been tested using a flowthrough system with a capacity of 300 m^3 h^{-1}. The actual volume of treatment in the system was 2 m^3 with an active acoustic area of 2 m^2. To inactivate a large number of real types of plankton such as nauplii, copepods, and rotifers, a specific power of 0.05 kWh m^{-3} proved to be satisfactory.

A continuing problem in water treatment is the occurrence of algal blooms. Algae may be killed relatively easily on exposure to ultrasound. Low concentrations of pollutants produce very little attenuation of sound transmission thus permitting the use of higher-frequency ultrasound and consequently low power emission and

Table 5. The Effect of Ultrasound and Chlorination on Bacterial growth[a]

Treatment	% Bacteria Killed after 5 min	% Bacteria Killed after 20 min
No treatment	0	0
Chlorine 1 ppm	43	86
Ultrasound alone	19	49
Ultrasound + chlorine	86	100

Note: [a]Conditions: 1:10 dilution of raw stream water, using ultrasonic bath (power input to system = 0.6 W cm^{-2}), T = 20 °C.

consumption. High frequencies give maximum activity at the interface between liquids and gases, as shown in sonoluminescence studies. Logically, then, if a large number of small bubbles were introduced into a field of high-frequency ultrasound, there would be a very large gas/liquid surface area for cavitational activity and the bubbles themselves should also provide "seeds" for cavitation events. This is the basis of an approach to alga removal and control proposed by the Belgian company Undatim [54]. In a trial involving the monocellular algal species *Scenedesmus capricornutum* some spectacular results were obtained. A reactor was constructed to treat water at a rate of 3600 L h^{-1} using an acoustic power of 175 W. At a temperature of 25 °C a solution of the algae containing some 100 algal cells per cm^3 was passed through the reactor, which reduced the recovery threshold of the microorganism by some 60%. This indicates that the treatment, operating at algal concentrations that are similar to those that might be encountered normally, offers the potential not only to kill the microorganism but also to severely restrict its reproductive ability. Similar results were obtained with more highly concentrated algal solutions.

One major question arises from such results, namely, what is the mode of action of the ultrasound? There is no trace of peroxide in the water after this form of treatment, thus eliminating cavitation as a possible source of destruction. Indeed at the powers used it is extremely unlikely that cavitation in the bulk liquid could be a major contributor to the effect. The inventors suggest that the algal cells themselves are activated in a manner similar to that proposed for the activation of cancerous cells in sonodynamic therapy [55]. In this technique the cancerous cells are killed after they selectively absorb molecules such as porphyrins, which, under the influence of ultrasound, generate active species such as singlet oxygen that kill the cell. In the algal cells it is suggested that the chlorophyll present may act in the same manner as the chemotherapy agents on cancerous animal cells.

This ultrasonic antialgal methodology has been combined with an electromagnetic antiscaling treatment to provide a new global water remediation technology for half-closed circuits, e.g., cooling towers, known as Sonoxide [56]. This process tackles two major problems of cooling circuits, namely the buildup of scale and algae. These are solved with a minimal energy requirement, without the need to use soft water and without the addition of chemicals.

The biocidal action of the ultrasound has been explained above. The effect of the electromagnetic field together with ultrasonic treatment provides the antiscaling properties. This combination seeds and orients the crystallization of calcium carbonate from hard water into nonscaling crystalline species. These species include vaterite and calcite, without aragonite, i.e., the same as would be obtained if chemical crystallization modifiers were used but in this case produced by purely physical action. The crystals formed are permanently carried away with the flow thus providing continuous cleaning of the cooling circuit. The sludge generated by the various species of calcium carbonate and from other impurities in the water, are separated by a hydrocyclone. Draining the cyclone takes only a few liters per day thus reducing the bleed-off to minimal quantities.

An installation of this type of equipment has been used for more than a year in the treatment of water in a cooling tower at the European headquarters of Dow Corning in Belgium. The cooling tower operates on a 550-kW cooling group with an identical tower nearby serving as reference. There is a major saving in water (for each evaporated cubic meter the Sonoxide tower drained 50 liters whereas the reference tower drained 2 m^3), the inlet water does not have to be softened by ion exchange, and the drain-off water is noncorrosive. Other advantages include:

1. The technology does not need chemicals.
2. After more than 12 months' operation with hard water, the heat exchanger did not show any traces of scaling or corrosion.
3. A bacterial test has given comparable, and even slightly better results than the reference tower.
4. No algae have developed, despite the complete absence of algicides.

One other demonstration of the efficiency of the process has been quoted by the inventors as evidence for the control of bacterial growth. A sample of 300 L of city distribution water was kept in a vat fitted with a Sonoxide side circulation loop. The water was maintained for 6 weeks at 40 °C with constant aeration and the bactericidal efficiency of the process was proven even under these conditions favoring bacterial growth.

Over the last decade there has been considerable interest shown in the use of ultrasound for water disinfection. The so-called "Sonozone" process was developed in Denmark in the early 1980s for the disinfection of water [57]. Inactivation of 3 to 4 decades of bacteria was obtained using from 10 to 95% lower ozone concentration and 57 to 96% lower gaseous ozone dosage by the sonozonation process compared with ozonation alone. This synergy was obtained by the simple fact that a gas, like ozone, can be diffused in minute microbubbles creating a high gas–liquid surface contact area allowing a more efficient yield of the treatment compared with ordinary aerators or ozonizers. Ultrasound also disaggregates the viable microbial units (*vide supra*).

TiO_2 suspensions in water were irradiated to inactivate *E. coli* and *Hansenula polymorpha* [58]. Two types of batch reactors employing static and recirculating solutions were used in the study. Sonolysis using a 20-kHz ultrasonic unit was found to enhance the microorganism inactivation in all instances, although the enhancement was more modest for the batch recirculation reactor. The mechanism based on sonolytic creation of ·OH radicals appeared to provide the most satisfactory explanation of the data trends observed. The data also implicated ·OH as the dominant bactericidal agent in irradiated TiO_2 suspensions.

6.2 Digestion of Sewage Sludge

Anaerobic fermentation is the most commonly applied process for stabilization of sewage sludge. Mass reduction, methane production, and improved dewatering properties of the fermented sludge are important features of anaerobic digestion. Because of carbon removal in the form of methane and carbon dioxide, the end product shows a substantially better biological stability than the unfermented material. Therefore, disposal on a dumping site or application as a fertilizer is possible. A disadvantage of the fermentation technique is the slow degradation rate of sewage sludge. Conventional residence times in anaerobic digesters are about 20 days, requiring large digestors.

Ultrasound has been used to accelerate the anaerobic digestion of sewage sludge [59]. The slow degradation rate of sewage sludge in anaerobic digestors is due to the rate-limiting step of sludge hydrolysis. Ultrasound treatment at a frequency of 31 kHz and high acoustic intensifies resulted in raw sludge disintegration shown by an increase in chemical oxygen demand in the sludge supernatant and a size reduction of the sludge solids. Semicontinuous fermentation experiments with disintegrated and untreated sludge were carried out for 4 months on a half-technical scale. The fermentation of disintegrated sludge was stable even at the shortest residence time of 8 days with biogas production 2.2 times that of a control fermenter. Ultrasonic treatment provided a better degradability of raw sludge, permitting a substantial increase in throughput.

6.3 Sterilization of Food

Heat treatment is one of the most frequently used methods for stabilizing foods because of its ability to inactivate enzymes and destroy microorganisms. However, since heat can also alter many organoleptic properties and diminish the content or availability of some nutrients, there is a growing interest in searching for methods that are able to stabilize foods with little or no heat added. The first report on the synergy between ultrasound and heat as a mechanism for killing the vegetative bacterium *S. aureus* was published by a Spanish group who found that the use of power ultrasound allowed a reduction in the effective temperatures at which sterilization could be achieved [60]. *Thermosonication* is the term now given to the combined application of heat and ultrasound and it was found to reduce the concentration of *Bacillus subtilis* spores by up to 99.9% in the 70 to 90 °C range in a small-scale ultrasonic reactor using a 20-kHz, 150-W ultrasound source [61]. Work carried out at Coventry University has addressed the issues of the effect of the food substrate (orange juice, milk, and rice pudding) on the thermosonication phenomenon using a range of organisms (*Zygosaccharomyces bailli*, *Listeria monocytogenes*, and *Bacillus polymyxa*) [62]. The studies confirmed the synergistic effect of ultrasound and heat; in milk, the heat resistance (D value) of *L. monocytogenes* was approximately 6-fold lower at 60 °C when sonicated at 20 kHz and the

D value of Z. *bailii* in orange juice was approximately 10-fold at 55 °C lower when sonicated at 38 kHz. Similar results were obtained in a study of the combined effect of ultrasound (20 kHz) and heat treatment on the survival of two strains of *B. subtilis* in distilled water, glycerol, and milk [63]. When spores suspended in water or milk were subjected to ultrasonic waves before heat treatment, little or no decrease of the heat resistance was observed. When heat and ultrasound were applied simultaneously, the heat treatment times in milk were reduced by 74% for *B. subtilis* var. niger-40 and by 63% for *B. subtilis* var. ATCC 6051 and similar results were obtained in glycerol. In water, however, thermosonication was more marked reducing the heat resistance of the spores by up to 99.9% in the 70–95 °C range; this effect of thermosonication was slightly diminished to 75% as the temperature reached the boiling point of water.

6.4 Decontamination of Surfaces

The use of power ultrasound for surface cleaning is a long-established and efficient technology. Ultrasound is particularly effective in this type of decontamination because the cleaning action is induced by cavitational collapse on and near surfaces which will dislodge bacteria adhering to them. The particular advantage of ultrasonic cleaning in this context is that it can reach crevices that are not easily reached by conventional cleaning methods. Objects for cleaning can range from large crates used for food packaging and transportation to delicate surgical implements such as endoscopes.

The removal of bacteria from various surfaces is of great importance to the food industry and can be efficiently accomplished with the combined use of sonicated hot water containing biocidal detergents. Normally plastic poultry trays are washed using a mixture of hot water spray (75–80 °C) and detergent (1% Trayclean); this technique can achieve bacterial kill rates of up to 60% [64]. The authors showed that thermosonication alone (75 °C) destroyed between 55 and 68% of the bacteria present, but thermosonication in the presence of a detergent (1% Trayclean) gave a kill rate of greater than 80%. However, it should be noted that a 100% kill rate was not obtained under the conditions studied.

6.5 Ultrasonically Assisted Antibiotics

A synergistic effect has been reported in biocidal action between ultrasound (67 kHz) and the antibiotic gentamicin at levels and concentrations that individually did not reduce viability of bacteria [65]. This synergy reduced the viability by several orders of magnitude for cultures of *P. aeruginosa, E. coli, Staphylococcus epidermidis*, and *S. aureus*. Measurements of the bactericidal activity of gentamicin against *P. aeruginosa* and *E. coli* demonstrated that simultaneous application of 67-kHz ultrasound enhanced the effectiveness of the antibiotic. As the age of these cultures increased, the bacteria became more resistant to the effect of the antibiotic alone and the application of ultrasound appeared to reverse this resistance. The ultrasonic treatment-enhanced activity was not observed with cultures of gram-positive *S. epidermidis* and *S. aureus*. These results may have application in the treatment of bacterial biofilm infections on implant devices, where sequestered bacteria are usually more resistant to antibiotic therapy. Synergistic killing was observed to be a function of ultrasonic intensity [66]. Greatest killing (approximately 5 log reduction in viable population) was realized at full intensity (4.5 W cm^{-2}), and decreased with reductions in power density. At the lowest intensity used (10 mW cm^{-2}) there was no significant evidence of acoustically enhanced killing.

The effect of ultrasound frequency on enhanced killing of *P. aeruginosa* biofilms by an antibiotic on polyethylene substrate has been studied [67]. Biofilm viability was measured after exposure to 12 μg cm^{-3} gentamicin sulfate and at an ultrasonic intensity of 10 W cm^{-2} at frequencies of 70 kHz, 500 kHz, 2.25 MHz, and 10 MHz. The results indicated that a significantly greater fraction of the bacteria within the biofilm were killed by gentamicin when they were subjected to lower-frequency rather than higher-frequency ultrasound. Experiments have been carried out to determine whether this bioacoustic effect was caused by ultrasonically induced changes in the morphology of the biofilm (biofilm breakup or disruption) [68]. Such disruption would be undesirable in the possible ultrasonic treatment of implant infections. A frequency of 500 kHz and an intensity of 10 mW cm^{-2} were used on biofilms of *P. aeruginosa* aged for 24 h and these were then examined by confocal scanning laser microscopy (CSLM). The CSLM results showed that the biofilm is a partial monolayer of cells with occasional aggregates of cells, noncellular materials, and extracellular spaces. The structure of the biofilm was not changed when it was exposed to continuous ultrasound, which, under the same irradiation parameters, increased cell killing by nearly two orders of magnitude.

An *in vitro* study investigated the bactericidal effects of a dental ultrasonic descaler on bacterial biofilms using *Actinobacillus actinomycetemcomitans* and *Porphyromonas gingivalis* [69]. Suspensions of the bacteria were subjected to the vibrations of a Cavitron P1 insert for 2.5 and 5.0 min in an acoustically simulated model substrate. A 60% kill rate was achieved at a temperature of around 50 °C, which constituted an alternative treatment for bacterial biofilms.

6.6 Effect on DNA

The effect of low-intensity insonation of cultures of *E. coli* (substrain B/r) has been studied [9]. The two experimental endpoints were the monitoring of cell survival (as indicated by the ability to replicate and form colonies) and the direct assessment of damage to cellular DNA. In this system the DNA is present in the intact cell in a long circular chain, attached to the cell membrane only at a relatively small number of attachment sites, the integrity of which are crucial for cellular replication. The survival curve (Figure 7) shows an exponential behavior. The bacteria are seen to be fairly sensitive to ultrasound, but there is an intensity

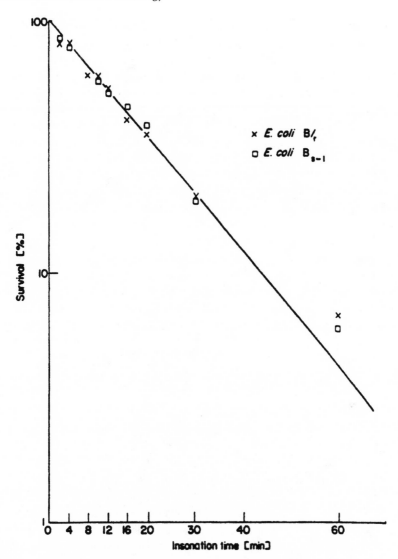

Figure 7. Postinsonation survival of two strains of bacteria. Experiments were carried out at an ultrasonic intensity of 3.5 W cm^{-2}.

threshold at a magnitude that indicates that the primary mechanism for bacterial kill is cavitation in the bulk medium and not within the cells themselves.

A technique has recently been developed to establish whether the DNA suffers any damage by assessment of strand breaks [70]. Results showed clearly (Figure 8) that in contrast to ionizing radiation, ultrasound does not damage intracellular

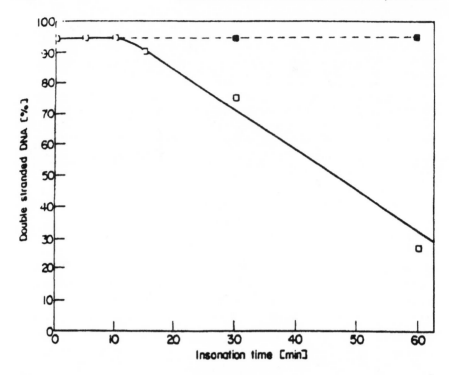

Figure 8. Effect of ultrasound on DNA after sonication at an intensity of 3.5 W cm^{-2}.

DNA at the intensities studied unless exposure times are greater than 15 min. This finding confirms the view held by many that experiments with DNA in solution, which can be degraded at very low intensities [9], have little relevance to the interpretation of ultrasound bioeffects *in vivo*. In addition, it has been observed that, at the intensities used, the bacteria themselves remained intact. It was concluded that shear forces set up by cavitation bubbles within the suspension medium act on the bacteria with insufficient force to rupture the cells, but they are sufficient to sever the more delicate attachment sites of the DNA to the membrane although the DNA chain itself remains intact. Prolonged sonication, however, does lead to strand breakage of the detached DNA chains.

7. OTHER APPLICATIONS OF ULTRASOUND TO MICROORGANISMS

Power ultrasound has the ability to deactivate enzymes through disruption caused by cavitation. Such deactivation might also occur within microorganisms, although there are some examples of the use of low-power ultrasound to enhance and activate microorganism metabolism. Thus, lactose hydrolysis and β-galactosidase activity have been increased by the use of ultrasound in the fermentation with *Lactobacillus* strains [71]. Milk fermentation using four *Lactobacillus* strains—*L. delbrueckii* ssp. *bulgaricus* B-5b, *L. helveticus* LH-17, *L. delbrueckii* ssp. *lactis* SBT-2080, and *L. acidophilus* SBT-2068—was carried out in the presence of ultrasound. Under continuous sonication, viable cell counts decreased compared with conventional fermentation, but if static incubation was used after a period of sonication the fermentation rate increased considerably. Sonication caused β-galactosidase release from lacto bacillus cells to the culture medium, thus resulting in higher total β-galactosidase activity. However, lactose hydrolysis was enhanced only when β-galactosidase was effectively released. With *L. delbrueckii* ssp. *bulgaricus* B-5b and *L. helveticus* LH-17, the degree of lactose hydrolysis achieved was about 75%, which is much higher than seen in conventional fermentation (below 40%).

The same group also investigated the effects of ultrasonic irradiation during milk fermentation in terms of cell viability, β-galactosidase activity, pH value of the culture medium, degree of lactose hydrolysis, and glucose content [72]. The results showed that the ultrasonic irradiation caused the intracellular β-galactosidase to be released from the lacto bacillus cells. The released β-galactosidase showed a higher lactose hydrolysis activity than that in the cells. High degrees of lactose hydrolysis and high cell viabilities were obtained with the combination of pH-controlled sonicated fermentation and static incubation.

8. CONCLUSION

The potential uses of power ultrasound continue to expand. From the foregoing it is proving to be a powerful tool for microbiologists to use not only in disinfection but also in cell separation, enzyme production, medicine, DNA manipulation, and improved fermentation. However, the use of ultrasound in microbiology is in its infancy but as more researchers make use of sonication the subject of sonomicrobiology will expand. It certainly offers great hope for the future of biotechnology.

GLOSSARY

Aerobe   An organism that requires oxygen for growth.

Algae   Simple photosynthetic plants.

Anaerobe   An organism that does not grow in the presence of air.

Antibiotic   A substance that is produced by microorganisms that in very low concentration inhibits or kills the growth of other microorganisms. Antimicrobial agents include synthetic compounds that have the same effect as antibiotics.

Antiseptic   A chemical that is applied to the body surfaces to prevent infection. This is achieved by inhibiting or killing microorganisms.

Asexual reproduction   Reproduction that does not involve the fusion of sex cells (gametes).

Autotroph   An organism that can assemble all of its organic components from inorganic matter.

Bacillus   A rod-shaped bacterium.

Bactericidal agent   An agent that can kill bacteria.

Bacterium   A prokaryotic microorganism.

Binary fission   Splitting into two parts. Binary fission of bacterial cells is the means by which bacterial populations grow.

Capsule   An envelope of carbohydrate or protein surrounding the cell wall of certain microorganisms.

Cell membrane   Controls the entry and exit of materials, allowing certain substances through, but prevents passage of others. It is a selective membrane.

Cell wall   A rigid structure, external to the cell membrane, that lends structure to plant, fungal, and bacterial cells.

Chemotroph   An organism that uses chemical compounds for its energy supply.

Cilium   A hairlike appendage to a cell that is capable of movement. With ciliate microorganisms, cilia may be used as a means of propulsion.

Coccus   A round-shaped bacterium.

Cyst   A type of spore formed by microorganisms.

Cytoplasm   All of the material that lies within the cell membrane. It is the liquid body of the cell in which all of the chemical reactions of life occur.

Cytosol   Liquid body of cell.

Debridement   Removal of debris/foreign particles from wound.

Denaturation   Alteration of nucleic acid or protein that results in the loss of normal biological activity.

Disinfectant   Chemical used for the decontamination of an environment that acts by killing live cells of microorganisms.

Disinfection   The removal or inhibition of microorganisms that are likely to cause disease from an object or environment.

Endospore   A thick-walled spore formed by microorganisms—see Spores. Endospores are highly resistant structures, affording protection against heat, irradiation, and chemicals. Often referred as *spores*.

Endotoxin   Toxin (lipopolysaccharide) associated with the cell membrane; leads to septic shock.

Eukaryote   An organism or cell having the cell nucleus separated from the cytoplasm by a membrane (nuclear membrane). The genetic material is borne on chromosomes consisting of DNA and protein. Cytoplasm consists of membrane-bound organelles.

Exotoxin   A toxic protein, produced by the normal metabolic processes of a microorganism and often released into its environment.

Facultative   An organism that grows in the presence or absence of oxygen, but grows better in the presence of oxygen.

Fermentation   The metabolism of organic compounds to release energy without the use of oxygen. Organic compounds are used as both electron donors and acceptors.

Flagellum   A thin whiplike appendage of microorganisms that is responsible for movement.

Fungus   A eukaryotic organism that possesses a cell wall, and that requires a supply of organic matter from which it derives energy.

Heterotroph   An organism that requires one or more organic compounds to grow.

Hypha   A filament that constitutes part of the fungal mycelium.

Lysis   The rupture of a cell.

Microorganism   Living organism too small to be seen by naked eye.

Organelle   A persistent structure with specialized function forming part of a cell.

Pathogen   An organism capable of causing disease.

Plankton   Microscopic animals (zooplankton) and plants (phytoplankton) that float in the surface waters of lakes and seas.

Prokaryote   An organism or cell having the genetic material in the form of simple filaments of DNA and not separated from the cytoplasm by a nuclear membrane. Many components or organelles found in higher organisms are missing.

Protist   The kingdom that contains mainly eukaryotic organisms.

Protozoa   Unicellular eukaryotic animal.

Spore   A resistant, dormant structure formed by microorganisms.

Toxin   A poison. In particular, toxins are substances produced by organisms that in low concentrations can damage other organisms and cause disease.

Trabeculated bone   Bone supported by fibrous tissue.

REFERENCES

[1] Mackean, D., and Jones, B. In *Introduction to Human and Social Biology.* John Murray, London, 1975.

[2] McKinney, R. E. *Microbiology.* McGraw–Hill, New York, 1960.

[3] Gaudy, A. F., and Gaudy, E. T. *Microbiology for Scientists and Engineers.* McGraw–Hill, New York, 1980.

[4] Cartwright, F. F. *Disease and History.* Hart–Davis, London, 1972.

[5] Hawker, E., and Linton, A. H. *Micro-organisms.* Edward Arnold, London, 1983.

[6] Beier, W., and Dorner, E. *Der Ultraschall in Biologie und Medizin.* Thieme, Stuttgart, 1954.

[7] Reid, I. M., and Sikov, M. R. *Interaction of Ultrasound and Biological Tissues.* Proc. Workshop, Washington, 1972.

[8] Sinistera, J. V. *Ultrasonics*, 30 (1992) 180.

[9] Graham, E., Hedges, M., Leeman, S., and Vaughan, P. *Ultrasonics*, (1980) 224.

[10] Leeman, S., Khokhar, M. T., and Oliver, R. *Br. J. Radiol.*, 48 (1975) 954.

[11] Glazunova, A. V., and Efimova, S. A. *Acoust. Phys.*, 43 (1997) 45.

[12] Lehninger, A. In *Biochemistry.* Worth, New York, 1975.

[13] Hammond, S. M., Lambert, P. A., and Rycroft, A. N. *The Bacterial Cell Surface*. Croom Helm, London, 1985.
[14] DeRobertis, E. D. P., and DeRobertis, E. M. F. *Cell and Molecular Biology*. Holt–Saunders, Tokyo, 1980.
[15] Griffin, J. E., and Touchstone, J. C. *Am. J. Phys. Med.*, 42 (1963) 77.
[16] Julian, T. N., and Zentner, G. M. *J. Pharm. Pharmacol.*, 38 (1986) 871.
[17] Phull, S. S., and Mason, T. J. *Confidential Report for Yorkshire Water*, 1995.
[18] European Patent EP 0 567 225 A1 (1993) for Biwater.
[19] Reynolds, C., and Wills, C. D. *Int. J. Radiat. Biol.*, 25 (1974) 113.
[20] Schram, C. J. *Advances in Sonochemistry*. JAI Press, London, 1991, Vol. 2, p. 293.
[21] U.S. Patent 5,164,094 (1992) and EP 0292470 (1987) for Stuckart.
[22] Coakley, W. T., Whitworth, G., Grundy, M., Gould, R. K., and Allman, R. *Bioseparation*, 4 (1994) 73.
[23] Peterson, S., Perkins, G., and Balter, C. *IEEE EM85 Annual Conference*, (1986) 4.
[24] Benes, E., Hager, F., Bolek, W., and Groschl, M. *Ultrasonics International '91, Conference Proceedings*, Butterworth–Heinemann, London, p. 167.
[25] Whitworth, G., Grundy, M. A., and Coakley, W. T. *Ultrasonics*, 29 (1991) 439.
[26] Miles, C. A., Morley, M. J., Hudson, W. R., and Mackay, B. M. *J. Appl. Bacteriol.*, 78 (1995), 47.
[27] Kowalska, E., and Mizera, J. *Ultrasonics*, 9 (1971) 81.
[28] Lorimer, J. P., Phull, S. S., Mason, T. J., and Pollet, B. *Electrochim. Acta*, 41 (1996) 2737.
[29] Baumgarte, F. *Arztliche Forsch.*, 3 (1949) 525.
[30] Mendez, J., Franklin, B., and Kollias, J. *Biomedica*, (1976) 121.
[31] Rapoport, N., Smirnov, A. I., Timoshin, A., Pratt, A. M., and Pitt, W. G. *Arch. Biochem. Biophys.*, 334 (1997) 114.
[32] Nyborg, W. L. *J. Cancer Res.*, 45 (1982) 156.
[33] Reisz, P. *Advances in Sonochemistry*. JAI Press, London, 1991, Vol. 2.
[34] Phull, S. S., Mason, T. J., Lorimer, J. P., Newman, A. P., and Pollet, B. *Ultrasonics*, 4 (1997) 157.
[35] Piczenik, Y., Kjer, B., and Fledelius, H. C. *Acta Ophthalmol. Scand.*, 75 (1997) 466.
[36] Betsch, J. M. *Recl. Med. Vet.*, 168 (1992) 1011.
[37] Jamni, L., Mdimagh, L., Jemnigharbi, H., Jemni, M., Kraiem, C., and Allegue, M. *J. Urol.*, 98 (1992) 228.
[38] Schwartz, J. H., and Ellison, E. C. *Postgrad. Med.*, 95 (1994) 149.
[39] McDonald, W. S., and Nichter, L. S. *Ann. Plast. Surg.*, 33 (1994) 142.
[40] West, B. R., Nichter, L. S., Halpern, D. E., Nimni, M. E., Cheung, D. T., and Zho, Z. Y. *Plast. Reconstr. Surg.*, 93 (1994) 1994.
[41] Zamani, A. F., Owen, R. W., and Clarke, D. J. *Soc. Appl. Bacteriol. Tech. Serv.*, 31 (1993) 55.
[42] Baker, N. V. *Nature*, 239 (1972) 389.
[43] Doblhoff-Dier, O., Gaida, T., Katinger, H., Burger, W., Groschl, M., and Benes, E. *Biotechnol. Prog.*, 10 (1994) 428.
[44] Hawkes, J. J., Limaye, M. S., Coakley, W. T., and Jenkins, P. *Ultrasonics World Congress 1995 Proceedings*, Gefau—WCU'95 Secretariat, c/o Gerherd-Mercator-Universitat, Duisburg, Germany.
[45] Hawkes, J. J., and Coakley, W. T. *J. Enzyme Microb. Technol.*, 19 (1996) 57.
[46] Hawkes, J. J., Limaye, M. S., and Coakley, W. T. *J. Appl. Microbiol.*, 82 (1997) 39.
[47] Kahl, J., Germer, R., and Ziegler, M. *J. Water Wastewater Res.*, 20 (1987) 38.
[48] Karwoski, J. P., Sunga, G. N., Kadam, S., and McAlpine, J. B. *J. Ind. Microbiol.*, 16 (1996) 230.
[49] Hughes, P., and Nyborg, N. *Science*, 138 (1962) 108.
[50] Alliger, H. *Am. Lab.*, 10 (1975) 75.
[51] Scherba, G., Weigel, R. M., and O'Brien, W. D. *Appl. Environ. Microbiol.*, 57 (1991) 2079.
[52] FWR Research Report. *Applications of Ultrasonics in the Water Industry*, 1995.

[53] Mues, A., Allied Signal ELAC Nautic. *Proceedings of 6th Meeting of European Society of Sonochemistry*, 1998, p. 53.

[54] Undatim Ultrasonics, Zoning Industriel, rue de l'industrie 3, B1400, Nivelles, Belgium.

[55] Umemura, S. I. *Jpn. J. Cancer Res.*, 81 (1990) 962, and subsequent papers.

[56] Cordemans, E., and Hannecart, B. WO 98/01394.

[57] Dahi, E., and Lund, E. *Ozone Sci. Eng.*, 2 (1980) 13.

[58] Stevenson, M., Bullock, K., Lin, W. Y., and Rajeshwar, K. *Res. Chem. Intermed.*, 23 (1997) 311.

[59] Tiehm, A., Nickel, K., and Neis, U. *Water Sci. Tech.*, 36 (1997) 121.

[60] Ordonez, P. *J. Dairy Res.*, 54 (1987) 61.

[61] Ordonez et al. *J. Appl. Bacteriol.*, 67 (1989) 619.

[62] Mason, T. J., Phull, S. S., Betts, G., and Earnshaw, R. *Internal Report (Confidential), Campden Foods*, 1995.

[63] Garcia, M. L., Burgos, J., Sanz, B., and Ordonez, J. A. *J. Appl. Bacteriol.*, 67 (1989) 619.

[64] Mason, T. J., Phull, S. S., Newman, A. P., and Charter, C. *Internal Report for Technopak plc*, 1994.

[65] Pitt, W. G., Qian, Z., and Sagers, R. D. *Ann. Biomed. Eng.*, 25 (1997) 69.

[66] Pitt, W. G., and Williams, R. G. *J. Biomater. Appl.*, 12 (1997) 20.

[67] Pitt, W. G., Qian, Z., and Sagers, R. D. *Colloids Surf. Biointerfaces*, 9 (1997) 239.

[68] Pitt, W. G., Qian, Z., and Stoodley, P. *Biomaterials*, 17 (1996) 1975.

[69] O'Leary, R., Sved, A. M., Davies, E. H., Leighton, T. G., Wilson, M., and Kiesser, J. B. *J. Clin. Periodontol.*, (1997) 24.

[70] Ahnstrom, G., and Edvardson, K. A. *Int. J. Radiat. Biol.*, 26 (1974) 493.

[71] Wang, D. Z., and Sakakibara, M. *Ultrasonics Sonochem.*, 4 (1997) 255.

[72] Wang, D. Z., Sakakibara, M., Kondoh, N., and Suzuki, K. *J. Chem. Tech. Biotech.*, 65 (1996) 86.

ULTRASONICALLY ASSISTED EXTRACTION OF BIOACTIVE PRINCIPLES FROM PLANTS AND THEIR CONSTITUENTS

Mircea Vinatoru, Maricela Toma, and
Timothy J. Mason

OUTLINE

Advances in Sonochemistry
Volume 5, pages 209–247.
Copyright © 1999 by JAI Press Inc.
All rights of reproduction in any form reserved.
ISBN: 0-7623-0331-X

Cleanse me with hyssop and I shall be clean.
Old Testament, Psalms, 51, 7.

1. GENERAL INTRODUCTION

The use of plants in daily life not only as food but also as flavorings, coloring, or in medicine, has a long history all over the world. The interest in aromatic and medicinal plants has declined over the last half-century, mainly due to the tremendous developments in the production of synthetic substitutes. Nowadays, there is a resurgence of interest in natural remedies, in part due to some disillusionment with modern medicine and drugs that either do not perform entirely to expectation or are accompanied by unwanted side effects.

Medicinal and aromatic plants provide an inexhaustible resource of raw materials for the pharmaceutical, cosmetic, and food industries and more recently in agriculture for pest control. People have learned to increase the power or usefulness of herbs, by preparing medicinal compounds from them, by preserving them so that they are always available, and by finding new ways to release their active constituents.

In this chapter we will explore solvent extraction as a method of obtaining the chemical constituents from plant material, and the development of ultrasonically assisted extraction to improve such procedures.

Ultrasonically assisted extraction of bioactive principles from plants and their constituents is not a new topic, but there is a dearth of information on systematic studies in this area. Recently the European Commission has granted us a Copernicus research program grant to study this topic, linking the United Kingdom (the Sonochemistry Centre at Coventry University), Romania, (the Costin D. Nenitzescu Institute of Organic Chemistry in Bucharest), and Slovakia (the Academy of Science and the Institute of Chemistry at Comenius University in Bratislava and the Mediplant company in Modra). Our goal was to develop ultrasonic methodologies and technologies to obtain plant extracts on a laboratory and pilot scale.

The main aim of this chapter is to provide a general overview of ultrasonically assisted plant extraction as well as some details of the experimental conditions employed. It is obvious that the Western world is now leaning toward natural products as a source of medicine and that there is an increasing demand for them. This is not a backward trend because over many centuries experience has taught us that we can find solutions to many problem illnesses if we look properly at natural resources. Plants provide what appears to be an endless source of benign chemicals for human needs. The development of new extraction techniques such as those involving ultrasound should encourage chemists to continue with their search for new bioactive constituents since the yields are likely to be improved. Once isolated, such materials could be used in their natural state or serve as model structures for the synthetic drug industry.

1.1 Historical Uses of Plants and Plant Extracts

One of the most ancient uses of plants was for nutrition, initially from natural and later from cultivated flora. Wheat, sunflower, corn, potato, beet, sugarcane, and many other plants are now cultivated on an extremely large scale for food and also as a resource for the chemical industry. On a smaller scale, saffron, obtained from the crocus, is used as a colorant and for flavoring in cooking. Some plants, such as lavender, chamomile, hyssop, and thyme, are cultivated for their essential oils commonly used in the cosmetic industry. Many other plants originating from wild, naturally occurring flora have also been used for the extraction of essential oils and other chemical components.

It is also possible that the isolation of a plant component can lead to the discovery of an important drug. One such example is aspirin, which was initially discovered in willow bark and leaves; indeed Hippocrates (400 BC) prescribed extracts of willow leaves to relieve labor pains during childbirth. Once its structure was elucidated and its synthesis optimized, this compound (salicylic acid) became one of the most used industrial manufactured drugs (as acetylsalicylic acid). Morphine and other alkaloids used in pain control are compounds that also originated as plant extracts. A recent discovery of the anticancer activity of another plant compound— taxol, produced by yew tree bark—illustrates the ever-expanding range of naturally occurring chemicals. Nowadays, industry is also searching for natural substitutes for fossil fuels and charcoal-based chemicals. The use of vegetable grease and tallow to produce biofuels and glycerol (as a by-product) is now a well-established industry and perhaps will continue to develop. This is a natural and renewable source of raw material to produce fuels, which is clearly not the case with mineral oil or coal derived from fossil sources. An old technique to produce natural ethyl alcohol by the fermentation of carbohydrate-containing plants has been reintroduced into industry. Ethanol is used not only for alcoholic drinks, but also as an alternative to fuel oil, and this natural source can compete economically with synthetic chemical processes. It is important to note here that many industrial technologies that use vegetable materials as feedstock do not produce environmentally unfriendly waste products.

The use of plants from natural and cultivated sources is an ancient human tradition for the treatment of wounds and combating disease. After many years of empirical experiment and accumulated knowledge, modern chemical and medicinal research on the potential uses of plant components continues apace with attempts to find the best procedures for extraction and application. The history of essential oils began in the East, especially in Egypt, Persia, and India. In the Egyptian papyruses there are thousands of recipes, showing that coriander and the castor oil plant were used for medicinal applications and essential oils as cosmetics and preservatives. Hebrew and Chinese manuscripts also describe over 2000 plants, offering details that remain useful today. During the Greek and Roman Empires the therapeutic use of plants was an expanding process. Hippocrates, the father of medicine, gives with full detail, advice on the use of 236 medicinal plants for hygienic and prophylactic purposes. He prescribed the treatment of an ill person as follows: first use psychotherapy, then phytotherapy, and only when these therapies fail resort to surgery. In ancient countries of the Orient and in ancient Greece and Rome an extensive trade was carried on in odoriferous oils and ointments. Modern history of plant extracts was begun by Galenus in the second century with around 30 papers and recipes, which are still up to date [1]. The first alcoholic perfume, known since 1380 as "Hungarian Water," was based on rosemary extract. For five centuries this was extensively used in Europe. In the 16th century the Swiss physician Paracelsus made an important contribution to the study of aromatic plants and their oils by using hot baths containing oils as treatments. The French school, in the 19th century, deepened the research in the field of plant products by setting up procedures to obtain desired extracts. Such extracts from the therapeutic plant species were known as "palace secrets" until the appearance of the first apothecaries.

In Romania the use of plants for curing disease has been known since antiquity. There is no firm evidence about when production of essential oils in Romania began. However, in the 19th century vegetable products were introduced into the Romanian pharmacopoeia. In 1904 the first Institute of Medicinal Plants in the world was established in Cluj-Romania [1]. Romania is a country with a temperate climate and a wide geographical variety providing very good conditions for the development of a rich natural flora. From over 3400 species of plants growing in this country, over 700 are considered as good sources of medicinal and aromatic compounds [1]. There is also a thriving agriculture based on the cultivation of certain plant species. Popular medicine in Romania has used many plant products such as extracts using boiling water, examples being tea, aromatic vinegar, and wines. Many of them passed into scientific medicine. Thus, "Coltea" hospital, Bucharest, founded in 1695, included a pharmacy selling medicinal herbs. The first Romanian Pharmacopoeia appeared in 1862 and included an important number of pharmaceutical extracts.

As early as the 17th century apothecaries were using licorice extracts to treat "inflamed stomachs" just as the ancient Chinese had done centuries before. Interest in plant-derived antiulcer drugs declined following the success of synthetic antihistamine drugs in the 1970s. Now interest has been revived with the identification of new compounds in plants, previously little known to Western scientists. With these new molecules from nature, scientists hope to combat more effectively one of the most uncomfortable and vexing problems in medicine [2].

Some years ago, Professor Harold Blum, from Oxford University, said: "all synthetic compounds are not compatible with the human body and therefore have to be accepted prudently by physicians as well as by the sick person; these substances must not be considered as harmless without the results of laboratory and *in vivo* tests" [3]. By comparison, natural compounds, obtained by extraction, are advantageous in that they have been proven useful because they have passed the test of time.

A number of plants have been investigated for their ability to control pests. The success of this is proved by the fact that some plant compounds, e.g., pyrethrum, have entered the industrial sector [4]. At this time it is increasingly important to apply horticultural technologies to get extracts of higher and reproducible quality. This is because plants, as a source of chemicals, are strongly influenced by soil and climatic conditions.

1.2 Classical Methods of Extraction

Classical extraction procedures could be classified in three main groups: distillation, solvent extraction, and compression to release oils.

1.2.1 Distillation

Steam distillation is the main procedure for the production of essential oils and employs water vapor at, or slightly over, atmospheric pressure. The distillate containing the volatile oils and water is cooled in order to separate the oil.

This method can be subdivided into three main procedures:

1. Direct distillation of essential oils
2. Water steam distillation
3. Water and steam distillation

Direct steam distillation involves the introduction of live steam (sometimes overheated) into a reactor containing the plant material (usually supported on trays). This procedure is suitable for all types of plant material except powders. It offers good penetration (especially when wet steam is employed), high rate of distillation, good oil yield, low hydrolyses, but is an expensive method. Sometimes it is necessary to use steam at higher pressure to achieve a good yield. A larger pressure means also a higher temperature that can cause cell membrane rupture or an enlargement of the pores thus opening new passages for the steam.

Water distillation of essential oils does not involve external steam. The crude plant material is heated in a closed reactor and water and volatile materials are distilled. This is a little-used procedure because of its inherent difficulties: nonuniform heating of plant material, overheating near the reactor walls, difficult temperature control, low penetration, degradation of plant components, low extraction rate, and so on. The same name is utilized for a technique that involves heating the vegetable material completely immersed in water. This technique has similar disadvantages.

Water and steam distillation of essential oils means heating the plant material (usually supported on a rack or perforated grid) with a predetermined amount of water. This leads to a water oil mixture that is later separated, the water being returned into the reactor. The disadvantages mentioned for water distillation apply here. However, the temperature control is improved as well as the oil yield, but some hydrolyses of esters take place.

There is no indication at present that ultrasound has been used in connection with these techniques, but the future development of ultrasonic devices leads us to conclude that it will be possible soon to introduce ultrasound during distillation to improve the yield of oil.

1.2.2 Solvent Extraction

In the past a very wide range of chemicals were used for solvent extraction. Today only a few organic solvents are in use for this purpose, mainly ethanol, ethyl acetate, glycerol, propylene glycol, petroleum ether with boiling point up to 80 °C. Other organic solvents are employed much less frequently or are rejected because of their toxicity. In the solvent extraction procedure plant materials are often chopped or milled first to expose the plant cells as much as possible to solvent contact. The techniques can be subdivided as follows:

1. Solvent (water or organic solvents) extraction of plant components
2. Maceration with water or alcohol–water mixture
3. Boiling with water (infusion)
4. Extraction with cold fat (enfleurage)
5. Extraction (maceration) with hot fat, etc. [1,5]
6. Liquid carbon dioxide extraction (including supercritical state)

In *solvent extraction*, the solvent is allowed to flow slowly over the plant material, placed in an appropriate vessel. The enriched solvent is then distilled and reused, leaving a semisolid extract containing essential oils, greases, waxes, proteins, and pigments. In the case of aromatic species the resulting residue is solid. By shaking this with alcohol to precipitate and eliminate the insoluble compounds, a very high quality extract is obtained. Usually this method needs additional agitation during extraction, and is suitable for ultrasonic treatment.

Maceration involves the solvent–plant mixture being allowed to stand several days (usually 1 or 2 weeks) at room temperature with occasional agitation. Ultrasound could be successfully employed during maceration to improve solvent penetration as well as to shorten the maceration time. This procedure leads mainly to tinctures.

Infusion entails the preparation of a plant–water extract using boiling water. Essentially making an infusion is similar to making tea. Heat and moisture readily cause the plant tissue to swell, the cell walls to expand, and the hydrodiffusion of plant constituents through the swollen membranes.

Enfleurage employs maceration with cold fat (from tallow or lard) followed by an appropriate solvent desorption (usually ethanol). This method is not widely used. Maceration with hot fat, essentially similar to enfleurage, gives a rapid maceration and consequently the time required is reduced, due to the high-temperature process.

Percolation is another solvent extraction procedure that uses different proportions of ethyl alcohol and water as solvent. This technique requires a percolator, a glass or metal column containing the vegetable material. The solvent, which is gradually added to the top of the column, flows slowly downward through the crushed plant material and extracts the soluble plant components. Percolation is connected with processes of osmosis and diffusion. At the end of this extraction procedure tinctures are obtained [5].

Liquid or supercritical carbon dioxide extraction is a new trend of plant material extraction. One of the first examples dealing with the extraction of hops by carbon dioxide in a supercritical state was published 20 years ago. Now this process is employed on an industrial level, especially for hop and coffee products [6,7].

1.2.3 Cold Compression

This procedure is used mainly in the natural oil industry to obtain vegetable oil from plants rich in oil such as sunflower, linseed, castor, and rape as well as for citrus fruit peels. The oil is literally squeezed out of the plant material. Sometimes ultrasound can be used during compression to increase the yield (some examples are given later).

Obtaining a plant extract is not a very easy task since several requirements have to be fulfilled. It is necessary to have a good knowledge of the crop horticulture, i.e., growing conditions and best harvesting time. Also, postharvesting handling, processing before extraction, suitable extraction procedures, and a method for purification of the extract are required. Many of these problems are the concern of farmers, but some of them are of interest for those who are doing the extraction. It is our opinion that the best way to accomplish this is to grow and process the plant material (including extraction) in the same region so as to form a production unit.

1.3   The Possible Benefits of Ultrasound in Extraction

Classical extraction procedures can be improved by the use of mechanical stirring, to increase the diffusion rate and to increase the surface of solvent–plant material contact [8,9]. Ultrasound can improve several aspects of an extraction process mainly through the phenomenon of cavitation. These improvements include the following.

1.3.1   Mass Transfer Intensification

One of the known mechanical effects of ultrasound is improved mass transfer arising from the collapse of cavitation bubbles at or near walls or interfaces. This leads to microparticle formation and emulsification, which creates a larger surface available to the solvent and microdispersion of one solvent into the other.

1.3.2   Cell Disruption

Ultrasound can break down the plant cell walls, releasing their contents. Moreover, by the milling effect on the bulk plant material, much more plant surface is in contact with the extracting solvent and, therefore, a better extraction of plant components can occur. Dispersion of the material and particle size reduction are other possible effects of ultrasound that may improve extraction.

1.3.3   Improved Penetration

Ultrasound is able to promote a better penetration of solvent into the plant cells and a resulting benefit in extraction yield. When cavitational bubbles collapse near the cell walls an ultrasonic jet is produced and drives the solvent toward the walls acting as a solvent micropump. This forces solvent into the cell dissolving the components and transporting them outside.

1.3.4   Capillary Effects of Ultrasound

The sonocapillary effect, or anomalous rise of a liquid in a capillary tube under the action of ultrasound, was discovered in the 1960s, but has been little investigated and poorly explained. Since some components such as essential oils are in the capillary system of plant material it is expected that ultrasound will assist by expelling them via this effect [10]. Some electrical phenomena accompanying the sonocapillary effect were reported by Ueda et al. in the 1950s (U-*effect* I and U-*effect* II) [11]. The authors showed that at both ends of a capillary an alternating electrical field is developed when ultrasound was applied to a capillary glass containing an electrolytic solution. This phenomenon could improve extraction of polar or ionic compounds, in a similar way to electrophoresis.

1.4   At What Stage of Extraction Can Ultrasound Be Used?

To see where ultrasound could be used to help extraction, it is necessary to establish the general unit scheme for each type of classical extraction. The general unit scheme for the water steam distillation technique to produce essential oils is shown in Figure 1.

Usually, the plant material and water are mixed into a distillation unit and heated together to the boiling point of water, to generate water vapor which codistills together with the essential oil. The oil–water vapor mixture is cooled and separated, and the water is returned into the distillation unit. By this method a good yield of oil is obtained, but at the end of the process a large amount of water containing soluble plant material remains and is a potential source of pollution on discharge. The steam distillation procedure is essentially the same as water steam distillation, the difference being in the production of the water vapor (see Section 1.2). After the vapor–oil mixture distillate is cooled and separated, the separated water could be reused to generate vapor and be reintroduced into the distillation unit, as this would produce less wastewater and thereby reduce pollution.

Sometimes stirring is used to improve the yield of essential oils during distillation, and in this respect one might imagine that an efficient ultrasonic technique could be employed instead of mechanical agitation to improve yields.

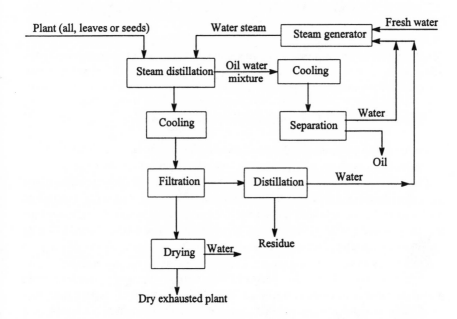

Figure 1. General operation unit scheme for water steam distillation.

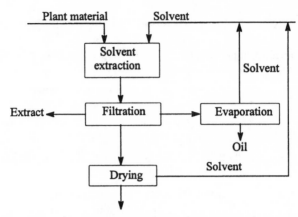

Figure 2. General operation unit scheme for organic solvent extraction.

Solvent extraction is usually performed at room temperature and so the logical place for an ultrasonic device is in the solvent extraction unit. In the case of aqueous or alcoholic solvent this may be an ultrasonic cleaning bath or a closed reactor fitted with a horn transducer. The latter type of unit could also be employed for volatile solvents like petroleum ether. The general operation unit scheme for solvent extraction is given in Figure 2.

2. THE DEVELOPMENT OF ULTRASONICALLY ASSISTED EXTRACTION

Among newer techniques used in extraction technology, the ultrasonically assisted extraction of oils [12] and other plant components has been employed as a new tool to improve the yield and quality of extraction products [13–16].

2.1 Essential Oils, Fats, and Resins

Among the earliest uses of ultrasound was in the extraction of fats and oils from vegetable sources. Nearly 50 years ago vegetable materials to be extracted were treated with a frequency of 20 MHz between electrodes in order to destroy the activity of lipase in materials such as rice bran [17]. This treatment had the effect of changing the character of the material to a more extractable form. Subsequently the material and a solvent were placed in an extractor and a frequency of 25 kHz was used to shorten the time of extraction.

It is perhaps worthy of note that not all applications of ultrasound in oil extraction result in improved yields. One such example is oil extraction by naphtha from pulverized and compressed sunflower cake [18]. A series of experiments was conducted on press cake granules on a pilot plant scale. Using frequencies of 19.6 and 21.3 kHz with ultrasonic intensities of 17 and 15 W cm^{-2}, and then at 300, 500, and 700 kHz at 20–25 W cm^{-2} there was only a small improvement in yield. Based on this result alone it might be concluded that the effect of ultrasonic irradiation would be too small to be useful in common extraction processes, although this has proved to be incorrect. A number of examples support this.

Consider the case of the extraction of oil from crushed peanuts using hexane at 400 kHz between 6.5 and 62.3 W cm^{-2}. A comparison was made between ultrasound and mechanical agitation. The results indicated that the effect of 60 W cm^{-2} ultrasonic treatment was equivalent to that of a stirrer operating at 1200 rpm [19]. Ultrasound has also been used to improve the extraction of oil from olive paste [8]. Minced plant material (olive paste) is subjected to a kind of electrophoresis in the presence of mechanical aids such as vibrators or ultrasound and under such conditions the extraction yield and quality were improved [20].

It has been reported that the continuous manufacture of animal and vegetable fats and oils by extraction can be improved using ultrasonic treatment if the moisture content of the vegetable and animal products is increased by the addition of water [21]. The improvement can be ascribed to a reduction in the binding power between the cell parts and the fat or oil. The fat or oil itself in the liquid state serves as a coupling agent for the transfer of acoustic energy to the vegetable material in a sonic extractor. Alcoholic extraction of oil seed with the aid of ultrasound has been carried out at 26 kHz within a power range of 0–20 W cm^{-2} [22]. The rate of extraction depended strongly on the geometric arrangement of the ultrasonic source with respect to the solid–liquid interface. It has been shown that ultrasound is capable of enhancing the solvent extraction of oil from palm fibers [23]. At frequencies between 20 and 60 kHz for only 12 s, using hexane or dichloromethane as solvent and under an atmosphere of CO_2, some 60% of oil was separated from the fibers.

Sonication has proven useful for the improved extraction of essential oils and resin from fragrant herbs. The volatile compounds from such plants can be used in medicine, as a biocide, in perfumery, and as spices. Ultrasound has been used for the solvent extraction of material from essential oil-bearing roses [24]. The physical constants of products obtained by this method were similar to those of material extracted by conventional methods but with ultrasound the extraction is more rapid and efficient than the usual technique of enfleurage. The extraction of fresh lavender by ultrasound at 22 and 800 kHz also gives a better yield and shortens the procedure by 2–3.5 times without diminishing the yield or reducing the quality.

The extraction of oak moss (*Evernia prunastri*) in an ultrasonic field improves the yield of odoriferous resins useful in perfumery [25]. The studies led to the following conclusions:

1. At low ultrasound intensity and at 70 °C the aroma was improved, but the resinous product was dark.
2. At moderate ultrasound intensity and at a suitable temperature, the yield was increased and the product improved in aroma and color.
3. At high ultrasound intensity the yield was low and the aroma deteriorated.

In a separate study the parameters for an ultrasound generator used in this process were optimized in terms of temperature and exposure time [26].

More recent studies in general extraction technology have applied sonication in the presence of a hard ceramic material constituting 50% of the volume of the vessel [27]. The essential oils are prepared from a suspension of plant parts in water or a sound-conducting medium (i.e., propylene–glycol) by exposure to ultrasound. Thus, a 30% mixture of water with grains, seeds, or beans, 5% propylene glycol, and ca. 20% hard ceramic ultrasound activators was irradiated with ultrasound at 35–40 °C and distilled. The yield of essential oils was 25% greater than by a conventional method. The authors explain that hard ceramic acts as an induced vibrator which can mechanically break down cell walls. This results in a better penetration by solvent and consequent extraction of the desired material.

A remarkable influence of ultrasound which gives one of the long-established industrial applications is for the extraction of essential oil and resin from hop cones. Very small quantities of hop oil are employed in the flavoring of certain liqueurs and in some cosmetic compositions. The practice of using hops for the flavoring of beer can be traced back in history to early times when mashed raw materials including hops were subjected to a fermentation process in order to obtain beer. Dried hops were used in beer making but proved not to be beneficial to the quality of the beer because long storage of the dried hop harvest seemed to damage the flavor. Subsequently an organic solvent extraction of hops was employed to avoid the loss of beer quality. There are a number of examples of the use of ultrasound to enhance such extractions.

A German work describes an ultrasonic procedure that used ultrasound for the extraction of hops [28]. However, using an appropriate dose of ultrasound a significant separation of bitter acids from tannins was achieved with the suppression of undesirable oxidation. Another ultrasonic treatment of hops was made by suspension of the ultrasonic device in the mash kettle. In the preparation of such beers, 30–40% less hops were used compared with untreated beer [29]. This appears to be the first application of a submersible ultrasonic transducer in the food industry. Another study describes a double hop-extraction process using ultrasound of 800 kHz [30]. Although ultrasound enhances extraction of hop bitters, no improvement is shown in the resulting beer owing to the loss of hop oils under usual extraction conditions. Thus, while hop bitters dissolve best at 100 °C (or higher) at 800 kHz the hop oils are destroyed whereas at 20 °C the hop oils dissolve without destruction although the bitters and tannins remain practically undissolved. Hence, a two-stage process first at lower and then at higher temperature should provide a better net result. The method involves adding the hop residues, as soon as possible after the hot extraction, to the wort within 1 h after the beginning of boiling. Following this the hot extract is added 15–30 min before the knockout of the wort. The later the hot extract is added, the more intense is the bitter taste and the greater the saving in hop material. Later addition of the cold extract enhances the final aroma and taste of the beer. In another study three brews of beer were mashed with the same raw materials: the first in the usual way, the second in a patented double-extraction process with ultrasonic influence using 33% less hop, and the third in the double-extraction process but using only 15% less hops. The three brews varied little in quality, foam, strength, and delicacy of the bitter flavor [31].

Other references to hop extraction include (1) a British patent which reports that an irradiation time of 3–4 h resulted in a 62% saving in beer hops using a frequency of 800 kHz at 0.93 W cm^{-2} [32], (2) the extraction of 35–50% of oils as an emulsion by passing hops and water through an intense cavitation zone produced by a 500 W cm^{-2} ultrasonic generator [33], and (3) the extraction of bitter substances from hops at 16.4 kHz and 98 °C for 30 min using a volume sonication power of 70 W L^{-1} and a specific sonication power of 0.4 W cm^{-2} [34].

Studies on the effect of ultrasound on the extraction of the main components of sage (*Salvia officinalis*) were performed recently, showing that cineole, thujone, and borneol could be extracted better when sonicated with a probe system [16]. Around 60% of these components are extracted within 2 h at a temperature varying between 8 and 33 °C. The authors noted that when using an ultrasonic cleaning bath at 20 °C and mechanical agitation, the yield of extracted compounds rose to near completion (Table 1).

Different types of extraction were performed using petroleum ether as solvent. This procedure could be applied to obtain extract as well as essential oils (petroleum ether is easily removed) [14]. A comparison between ultrasonic and Soxhlet extraction methods of milled dill seeds in hexane is given in Table 2. The ultrasonically assisted extractions were performed in an ultrasonic cleaning bath, using indirect sonication.

It is interesting to note that at longer extraction times the concentration of limonene decreased even when ultrasound was used. This is quite strange since the total amount of oil extracted is greater than from classical or short-term ultrasonic extraction. It seems that maceration alone hinders the extraction of limonene but the reason for this is unknown. One of the most important observations is that

Table 1. Comparison of Different Extraction Procedures

Extraction Procedure	Component Extracted (mg/kg)		
	Cineole	Thujone	Borneol
Probe (2 h)	24.3	167.2	5.8
Cleaning bath[a]	16.5	118.1	3.6
Cleaning bath + stirrer[a]	22.7	141.9	6.3
Silent control experiment[b]	14.0	176.6	6.5

Notes: [a]For these procedures data collected at 3 h.
[b]Best silent results.

Table 2. Comparison between Different Extraction Methods of Dill Seeds, Using Hexane

Method Used[a]	Oil Amount (g)	Extraction Time (min)	Component		
			Limonene %	Carvone %	Heavy Components %
CE	3.00	240	40.79	47.29	0.09
US	3.40	30	49.63	48.15	—
US	3.40	60	51.22	45.84	—
M + US	3.50	7 days M + 30 US	25.08	66.84	0.05
M + US	3.50	7 days M + 60 US	20.77	65.40	0.34
M + US + R	3.55	7 days M + 30 US + 240 R	20.72	65.69	0.10
M + US + R	3.55	7 days M + 60 US + 240 R	20.28	65.92	0.34

Note: [a]CE, classical extraction (Soxhlet); US, ultrasonic extraction; M, maceration; R, reflux.

ultrasonically assisted extraction appears to reduce the amounts of heavy components to very low levels.

A comparison of classical silent and ultrasonic procedures for a series of plants using petroleum ether and ethanol (neat and aqueous) was published recently [15]. In the case of petroleum ether, Soxhlet, direct, and indirect sonication of dill seeds led to different amounts and different qualities of oil (Table 3). Direct sonication involved a closed reactor fitted with an ultrasonic horn transducer and mechanical stirrer or material plus solvent placed directly in the reactor. Indirect sonication used an Erlenmeyer flask containing the plant material and solvent immersed in an ultrasonic cleaning bath containing water as the coupling fluid.

Ultrasound improved the yield of oil (reported as grams of oil per 100 g seed) which was obtained in a substantially shorter time even in the case of indirect sonication. Significantly both types of sonication produced lower amounts of heavy components (mainly waxes). This is almost certainly the result of the shorter extraction times used in the ultrasonic processes. The most probable mechanism for the ultrasonic enhancement of extraction is the intensification of mass transfer

Table 3. Comparison between Classical and Ultrasonically Assisted Extraction of Dill Seeds

Method Used	Oil Amount (g)	Extraction Time (h)	Heavy Components (%)
Soxhlet	3.0	4.0	11.95
US (indirect)	3.4	0.5	3.14
US (direct)	3.6	0.5	3.06

and easier access of the solvent to the cell contents. Although cavitational collapse in petroleum ether does not produce high energies it would still seem to generate some cell disruption and good penetration of solvent based on the fact that direct sonication, which involves greater energy input, produces a greater yield of oil. In the classical procedure the mechanism is via normal diffusion through the cell walls—a process that requires substantially longer extraction times.

In the case of ethanol as solvent the plant material (dry and crushed) was sonicated using only the direct methods for 2 h (very few plant materials required more than 2 h). Both sonicated and control samples were allowed to stand (maturate) for 18–24 h before product isolation. A comparison of classical and sonic extraction procedures leading to tinctures is given in Table 4. The first six columns show the results of direct sonication (cleaning bath) for a series of plant materials, the last column gives the results for sonication by the probe system.

The amount of dry residue obtained from extracts depends strongly on the plant material. Thus, mint leaves showed only a small amount of extraction in the first half-hour whereas the other plants showed a rapid extraction within 15 min. This can be explained by the fact that mint leaf cells cannot be destroyed by a simple crushing process. Ultrasonic disruption of cell walls thus takes some time (ca. 30 min) after which the release of cell contents is much more rapid. In almost all of the ultrasonic cases the amount of extract is similar to or greater than the classical technique. The ultrasonic procedure thus seems to be a significant improvement when extraction time is taken into account. A similar trend in extraction was observed when the process was scaled up in a large cleaning bath (10 liters solvent and over 1000 g plant material).

Standard (silent) extraction (4 h, agitation, room temperature) was compared with the ultrasonically assisted method (0.5 h, agitation, room temperature) in 60% ethyl alcohol for milled seeds of coriander, fennel, and dill. The extracts were analyzed

Table 4. Dry Residue (g per 100 g Extract) Obtained by Direct Sonication

Sonication Time (min)	Mint[a]	Camomile[a]	Sage[a]	Arnica[a]	Gentian[a]	Marigold[a]	Marigold[b]
15	0.06	1.10	0.58	0.36	—	0.94	—
30	0.07	1.30	0.80	0.42	1.67	0.98	1.15
60	0.25	1.43	0.92	0.67	2.66	1.14	1.74
90	0.78	1.56	0.94	1.06	2.71	1.33	1.97
120	0.82	1.79	1.13	1.20	3.24	1.75	2.25
180	—	1.80	—	—	—	—	—
18 h maturation	0.91	1.91	1.15	1.50	4.68	2.20	2.25
Classical 7 + 14 days	1.02	1.73	1.02	1.75	4.75	2.25	2.25

Notes: [a]Cleaning bath.
[b]Probe system.

Table 5. Dry Residue and Grease Content Obtained by Direct Sonication Using a Probe System

Sonication Time (min)	Coriander		Fennel		Dill	
	Dry Residue %	Grease %	Dry Residue %	Grease %	Dry Residue %	Grease %
5	0.34	0.132	0.56	0.250	—	—
15	0.45	0.146	0.77	0.278	0.26	0.123
30	0.61	0.181	0.88	0.286	0.28	0.115
Classical	0.53	0.410	1.27	0.324	0.22	0.119

for dry residue and grease content. The latter estimation was accomplished by cooling the extract to between 0 and 4 °C to precipitate the grease so that it could be removed by filtration. The results are shown in Table 5.

Once again the use of ultrasound improves the extraction in all cases and this is accompanied by a reduction in the amount of grease. This can be attributed to an increase in the selectivity of extraction produced by ultrasound through (1) a shorter extraction time, (2) a reduction in the diffusion process which yields grease, and (3) an acceleration of the extraction process for low-molecular-weight compounds.

Vanilla beans were treated with ultrasonic waves of 300, 500, and 2000 kHz at 3 to 28 W cm^{-2} for periods of 1, 2, and 3 min [35]. The oleoresins were extracted from 5 g beans in 100 cm^3 ethanol at 50 °C for 12 h and the amount of weight loss of the beans determined after drying at 73 °C. The ultrasonic treatment had increased the vanillin extract content to 25–35% and the oleoresin extract content to 10–15%.

These selected examples serve to illustrate the effects of ultrasound especially for the extraction of desired compounds from raw plant materials. While ultrasound has been shown to be generally useful in improving extraction of oils it is also particularly good when used for the removal of residual oils remaining after conventional extraction has been performed.

2.2 Alkaloids

Several studies have been conducted on the ultrasonically assisted extraction of alkaloids from plant sources. In the classical maceration procedure plant material is first soaked in a suitable solvent or mixture of solvents which causes it to swell and this is followed by the extraction procedure which consists of the physical separation of the alkaloids from the plant, commonly by percolation or continuous extraction using a Soxhlet apparatus.

Among the range of alkaloids extracted from a wide variety of species, perhaps the most famous is the antimalarial medicine quinine extracted from cinchona bark (*Cinchona succirubra*). Schultz and Klotz [36] employed both high-frequency

Table 6. Effect of Ultrasound on Alkaloids Extracted by Maceration of Cinchona Bark

Sample	% Alkaloids Obtained after Different Sonication Times (min)			
	No Ultrasound	20	30	40
1	0.669	1.025	0.847	0.635
2	0.678	1.031	0.805	0.613
3	0.635	1.015	0.789	0.624

ultrasound (2.4 MHz) and sound in the audible range (less than 20 kHz) to determine the effect of ultrasound on this extraction. An improved yield was obtained only with the audible frequency. This confirms that cavitation is important in the process since it is unlikely to occur at the higher frequency used. An increased yield of cinchona alkaloids can be obtained using frequencies of 20 and 450 kHz [37]. Maceration and percolation techniques have been employed to determine the effectiveness on cinchona bark extraction of different periods of sonication [38]. Using 680 kHz and 35 W power generated from a transducer with an irradiating face area of 9 cm^2 the maximum yield was obtained with a sonication time of 20–25 min (Table 6). Results of the ultrasonic percolation procedure applied to the extraction of cinchona alkaloids are shown in Table 7.

The efficiency of a "step-horn" ultrasonic probe on the extraction of alkaloids has been investigated by Ovadia and Skauen using a 20-kHz ultrasonic generator under controlled temperature conditions [39]. The plant materials used were cinchona bark (*Cinchona succirubra*), ipecac root (*Cephaelis ipecacaunha*), and jaborandi leaf (*Pilocarpus micro-phyllus*). In the case of cinchona bark the filtrates were concentrated and assayed by methods described in *The National Formulary IX*.

It is evident from the data that a greater yield resulted from ultrasonic extraction without temperature control than by either Soxhlet extraction or temperature-controlled ultrasonic irradiation. Ultrasonic treatment for 45 min was nearly

Table 7. Effect of Ultrasound on Alkaloids Extracted by Percolation from Cinchona Bark

Sample	% Alkaloids for Different Sonication Times (min)					
	No Ultrasound	5	10	15	20	30
1	0.879	1.052	1.103	1.201	1.241	1.168
2	0.901	1.003	1.100	1.193	1.280	1.094
3	0.899	1.080	1.091	1.145	1.203	1.006

equivalent to Soxhlet extraction for 7 h. This compares favorably with results obtained by Head et al. [37] using a bath-type ultrasonic device. Ultrasonic extraction of ipecac root was conducted in duplicate by the methods shown above and the results are given in Table 8 [39].

The filtrates were concentrated and assayed by methods reported in the *U.S. Pharmacopoeia XVI*. In this instance, within 0.5 min, the alkaloid concentration produced by ultrasound was greater than that by Soxhlet extraction for 5 h. After some 5 min, complete extraction seemed to have taken place. Ultrasonic extraction of jaborandi leaf was conducted in duplicate with and without temperature control. The filtrate was assayed by methods reported in the *British Pharmaceutical Codex* (1949) and the results were compared with Soxhlet extraction. The data reveal that 15 s of ultrasonic treatment (20 kHz) extracted more alkaloid from jaborandi leaf than 5 h of Soxhlet extraction [40]. Cooling the insonated liquid mixture appeared to be beneficial since a higher yield was produced in 0.5 min possibly because the extracts seemed to degrade at a slower rate. In all three extractions, there was evidence of some degradation under the conditions of the experiments.

Another series of experiments investigated the effect of ultrasound on the extraction of the alkaloid belladonna from the leaves of the plant of that name (also called deadly nightshade, *Atropa belladonna*). Wray and Small carried out the irradiation of a maceration mixture for 20 min followed by Soxhlet extraction [41]. They reported the same yield of alkaloids as a conventional 8-h maceration followed by Soxhlet extraction. Using a frequency of 800 kHz and an intensity of 2.5 W cm^{-2}, 36.7% more alkaloids could be extracted compared with mechanical agitation

Table 8. Percent Total Alkaloids from Ipecac Extraction (Chloroform, Various Methods)

Extraction Time	US Extract	Temp.	US Extract Controlled Temp.	Temp.	Soxhlet Extract
0.25 min	0.79	27	0.785	20	—
0.5 min	0.89	29	0.900	20	—
1 min	0.91	31	0.920	20	—
3 min	0.94	41	0.955	23	—
5 min	0.96	46	0.960	25	—
15 min	0.96	52	0.950	25	—
0.5 h	0.95	55	0.965	25	0.214
1 h	0.96	55	0.950	25	0.380
1.5 h	0.96	55	0.960	25	—
3 h	0.95	55	0.960	25	0.710
5 h	—	—	—	—	0.840
7 h	—	—	—	—	0.830
Crude drug					0.830

employing maceration techniques. The yield was only 2.4% superior to percolation, but the time taken was shorter. The amount of alkaloid extracted reached a maximum after treatment for 15 min. No decomposition was observed and temperature did not appear to influence the extraction [42].

Shinyans'kii and co-workers studied the extraction of digitalis, belladonna, valerian, wild rose, and others, with application of ultrasonic waves (480–500 kHz and 20 W cm^{-2}) to the vegetable material in a suitable solvent. The procedure was repeated and the two extracts combined and aged. They noticed that the aging period was considerably shorter when ultrasound was used because of its coagulative action. Analyzing the extracts obtained by the action of ultrasound and by percolation, it was found that 10% more compounds were extracted during sonication compared with the silent method [43].

A maximum yield obtained by the percolative process, i.e., 11–18%, was obtained in 15 min at 50–60 °C without decomposition when ultrasound was applied in laboratory experiments to accelerate the extraction of alkaloids from belladonna leaves [44]. The apparatus used was not considered to be suitable for large batch operations.

Complete extraction of the alkaloids belladonna and henbane (from the plant of that name, *Hyoscyamus niger*) from crushed leaves and strychnine from crushed nux vomica seeds was achieved after ultrasound exposure for 30 min using two treatments with 5 and then 3 vol of solvent. Conventional percolation requires 12 vol of solvent and 21 h of treatment. The alkaloid content of the dry residue was extracted with the aid of ultrasound and compared with classical methods (Table 9). The alkaloid composition as determined by paper chromatography was found to be identical by both techniques [45]. The quantity of alkaloids obtained with the ultrasonic procedure does not differ significantly from that produced classically but both the time (from 21 to 0.5 h) and the solvent volume required are substantially reduced.

Morphine has been extracted from the dried leaves, seeds, pods, and stalks of the poppy plant mixed with H_2O (ratio 1:8) at 80–85 °C [43]. The yield obtained after exposure for 15–17 min to ultrasound of 500 kHz and 20 W cm^{-2} was found to be the same as that produced by classical extraction for 24 h.

Table 9. Comparison of Total Alkaloids Obtained from Three Sources with and without Ultrasound

Species	% of Alkaloids Extracted	
	Ultrasonically	*Classically*
Nux vomica	14.60	15.50
Herbane	0.33	0.35
Belladonna	1.47	1.38

An early study reported the extraction of morphine from the pods of opium poppies using ultrasound of 22, 300, and 750 kHz [46]. The amount extracted into the aqueous phase during a 10-min exposure at each frequency was essentially the same, namely, about 60% of morphine. Because morphine is degraded by sonication the yield of morphine was found to decrease with increasing time of exposure after the first few minutes. The yield was increased when an alcohol–$CHCl_3$–NH_3 mixture or aqueous HCl, $Ba(OH)_2$, or sulfamic acid solutions were used as extractants instead of pure H_2O. Aqueous $Ba(OH)_2$ (50 g/liter) extracted other alkaloids in addition to morphine but this did not occur with less concentrated solutions of $Ba(OH)_2$. Some years later, Aimukhamedova's group was able to significantly increase the yield of morphine obtained with ultrasonic extraction by using centrifugal activators at 78 °C [47]. Under these conditions the yield reached up to 97%.

DeMaggio and Lott [48] made a clear distinction between ultrasonic maceration and ultrasonic extraction procedures involved in the isolation of alkaloids from thorn apple (*Datura stramonium*). To study the influence of frequencies of 20 and 40 kHz during the maceration of the plant, the following procedure for ultrasonic maceration was employed: The recommended U.S. Pharmacopoeia quantities of plant and macerating solvent were placed in polyethylene containers and the containers were then immersed in the ultrasonic bath and sonicated for periods of 0.5 or 1 h. After the ultrasonic maceration was completed, the samples were

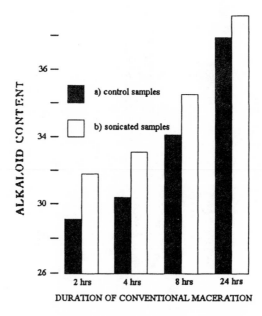

Figure 3. Results of 20-kHz ultrasonic extraction of alkaloids from thorn apple.

transferred to extraction thimbles and extracted for 3 h in a Soxhlet. It was found that the 0.5-h sonication at 20 kHz produced a 9% greater yield of alkaloids (Figure 3); sonication for 1 h did not increase the yield. The 40-kHz generator was demonstrated to be more effective than other units employed in releasing the alkaloids from the plants during the short maceration periods used (Figure 4).

The improvements brought about by sonication could well be the result of the physical ways in which the alkaloids are held within the plant. Some of the material will be present in the cell as salts of organic acids but a proportion may be present bound to essential cell constituents or precursors. The mechanism of the ultrasonic effect on maceration would be twofold, first to liberate the bound alkaloids and then to hasten their diffusion through the cell wall into the extracting solvent. In the absence of ultrasound, the liberation of the bound alkaloids and their diffusion occurs at a much slower rate. After longer periods of extraction, conventional extraction is able to remove most of the alkaloid and the effect of ultrasound became less significant. Therefore, short periods of ultrasonic maceration have been found to be more effective in liberating the desired alkaloids from plants. For ultrasonic extraction work, an apparatus has been developed that makes it possible to subject the plant to ultrasound while, at the same time, performing continuous extraction of the alkaloids. An experiment was carried out to determine the interaction of maceration time and extraction time: The plant was macerated for between 2 and 24 h and extracted for a fixed period of 2 h. The data suggested that ultrasonic

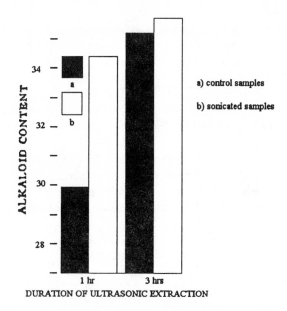

Figure 4. Results of 40-kHz ultrasonic extraction of alkaloids from thorn apple.

extraction was most efficient when employed in conjunction with short periods of conventional maceration. When a comparison was made between samples of the plant continually macerated for 24 h and extracted for 3 h with 20- or 40-kHz ultrasound, the yield of alkaloid was greater with the 20-kHz frequency. In this case the ultrasonic action causes no detectable degradation of the alkaloids. It was proposed that of the two effects of sonication, i.e., the liberation of bound alkaloids and then improvement in the diffusion process, the former was stimulated or enhanced by higher-frequency ultrasound, whereas the latter was thought to be promoted by low-frequency ultrasound. Ultrasound energy shortens considerably the duration of both the maceration and extraction processes but the maceration process lends itself most readily to the application of ultrasound. One of the most important conclusions of this work was that ultrasound definitely has a place in the extraction of alkaloid-containing plants and that ultrasonic treatment on a commercial scale could be utilized profitably and simply by applying ultrasound to the macerating mixture.

Rauwolfia bark (*Rauwolfia serpentina*) is a plant product known for thousands of years in ancient Indian medicine. It contains the alkaloid reserpine, which is now a well-known drug for the treatment of high blood pressure. A significant increase in the rate of extraction of this substance was observed in the initial stage of the exposure to ultrasound at 25 kHz. Reserpine obtained by the ultrasonically assisted method was chromatographically identical to the alkaloid obtained by normal methodology [49]. The maceration time of 8 h, according to the method described in the Indian Pharmacopoeia, was reduced to 15 min when 25-kHz ultrasound was applied [50].

Extraction of powdered nux vomica seed using ultrasound gave strychnine and brucine in 0.85 and 1.15% yields, respectively, within about 20 min compared with yields of 0.64 and 0.94% within about 8 h for the control [51]. This same extraction at 1 MHz and 3 W cm^{-2} in an oil bath gave a maximum yield in 30 min of brucine 1.2% and strychnine 0.95%. For industrial extraction, a flow system with 20–40 kHz and 3–5 W cm^{-2} is suggested with 25- to 100-mesh crushed seed. From the data, a 20-kW ultrasonic generator appears to be sufficient to produce the alkaloids at 50 kg/day [52].

More recent studies have applied sonication to improve the extraction rate of berberine. Comparison with alkaline immersion extraction, the ultrasonic extraction rate over 30 min is more than 50% that of alkaline impregnation over 24 h. The authors have shown that the use of ultrasonic technology is both simple and time saving [53].

2.3 Other Pharmaceuticals

Many species contain pharmaceutically active compounds other than alkaloids and the extraction of many of these has been exploited in order to obtain pharmaceuticals using ultrasound treatment.

Under appropriate conditions sonication does not degrade kellin, visnagin, and other furanochromones and so for this reason it has been used by the pharmaceutical industry to facilitate the extraction of furanochromones from the fruit of the *Ammi visnaga* plant. In a cleaning bath at 20 kHz, 2.2 W cm^{-2}, and 50–60 °C the extraction rate was improved five-fold over conventional methodology [54]. The optimum time of sonication was 30 min and this gave the maximum yield of furanochromones regardless of sample weight. Another method for the extraction of kellin has been proposed by Boeva and Dryanovska using 800 kHz at 10 W cm^{-2} [55]. Temperature has an important influence on this extraction and the results are shown in Table 10.

The optimum temperature for the experiment was established to be 80 °C. The maximum amount of kellin was extracted using 1:100 plant material:water after treatment for 45 min. A two-stage extraction is proposed by the authors whereby after the first 30 min of ultrasonic treatment, the extraction medium was replaced with fresh water in the same ratio. When the extraction was continued for another 15 min, the amount of kellin extracted was increased by 10%.

Alexandrian senna (*Cassia acutifolia*) contains anthraquinone glycoside (sennoside A and sennoside B) used as purgatives. The effects of ultrasonic energy on the aqueous extraction of senna pericarps were investigated by Morrison and Woodford employing 20-kHz ultrasound at an intensity of 60 W [56]. Whole pericarps were steeped in purified water for 3 min at room temperature and then subjected to ultrasonic treatment for 90 min. During this time the temperature of the extraction suspension rose from 20 °C to 50 °C. All experiments were done in triplicate and the results are shown in Figures 5 and 6.

Since the active constituents are susceptible to hydrolysis and oxidation processes, the effect of free radical formation during extraction was investigated. It was found that the extraction rates of sennosides and free anthraquinone were not affected by free radical formation. Thin-layer chromatography examination revealed no chemical degradation as a result of ultrasonic irradiation. An investigation was conducted to determine the effects of the more powerful ultrasonic horn system for the extraction of aglycons from *C. acutifolia*. The amount of aglycons extracted by this means was compared with aglycons extracted by a standard infusion method and the results are shown in Figure 7.

Table 10. Effect of Temperature on the Ultrasonic Extraction of Kellin

Extraction Method	Temperature °C	% kellin
Control	80	0.89
Ultrasonic	30	0.37
Ultrasonic	75	1.20
Ultrasonic	80	1.11
Ultrasonic	85	1.20

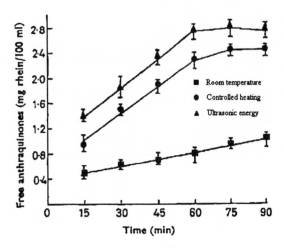

Figure 5. Effect of ultrasonic energy and controlled heating on the aqueous extraction of free anthraquinone from senna pericarps.

The amount of heat applied during the ultrasonic treatment has an important effect in the extraction of *C. acutifolia*. When boiling water was added followed immediately by sonication the amount of aglycons extracted was increased (Figure 8) [57]. The maximum extraction occurs after about 5 min with ultrasound versus about 10 min for the infusion method with nearly 17% more aglycons extracted by

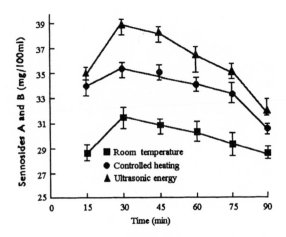

Figure 6. Effect of ultrasonic energy and controlled heating on the aqueous extraction of sennosides A and B from senna pericarps.

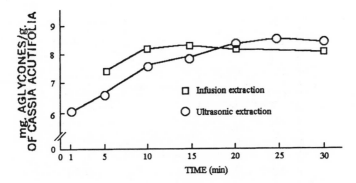

Figure 7. Comparison of ultrasonic extraction without boiling water and infusion extraction with boiling water.

sonication than by infusion. Over a 30-min period there is no significant amount of degradation by either ultrasound or boiling water.

The extraction of chlorophyll is commercially important because it is a large component in the production of many cosmetics. The crushed dried stinging nettle *Urtica dioica* was subjected to extraction under 1-MHz ultrasound and methanol was employed as the solvent. It was shown that the ultrasonic treatment of nettle results in the successful separation of chlorophyll depending on the ultrasound intensity and on the duration of exposure (Table 11) [58]. The extraction of chlorophyll was facilitated by ultrasonic treatment, since the ultrasonic energy transmitted to the chloroplasts leads to the freeing of chlorophyll from the vegetable material.

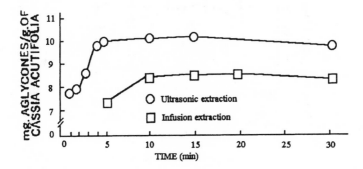

Figure 8. Comparison of ultrasonic extraction with boiling water and infusion extraction with boiling water.

Ultrasonic energy was used to shorten the extraction step for *Ephedra*. When applied to the powdered plant material in contact with alcohol and chloroform in a stainless-steel conical flask, ultrasonic energy shortened the maceration time to 20 min (instead of the usual overnight maceration and percolation) [59].

Extraction of sesquiterpene lactones from *Inula helenium* and *Telekia speciosa* was carried out with 95% ethanol using ultrasound of 800 kHz and 4.5 W cm^{-2} for periods between 10 and 40 min. After two filtrations the extracts were evaporated to dryness and the residue was subjected to purification on an alumina column. The amount of helenine extracted increased with increasing time of the ultrasonic treatment which did not affect the structure of the helenine components. The alantolactone/iso-alantolactone ratio in the helenine extract was 2:1 [60].

Romanian studies [61] were carried out using 1-MHz ultrasound in the extraction of amino acids from *Amaranthus retroflexus*. The only free amino acid detected after the extraction was glycine. Ultrasonic extraction was followed by ultrasonic hydrolysis of the proteins contained in the extract. Thirteen amino acids were identified by paper chromatography in the hydrolysate. In order to avoid chemical modification of amino acids under the ultrasonic irradiation, only short periods of exposure were employed. The results are shown in Table 12.

Results were calculated using the densitometric method and expressed in milligrams of amino acid per 100 g plant material. A 5-min exposure gave higher yields than 10 or 15 min, and so 5 min was recommended as the treatment time since further sonication appeared to result in product decomposition.

The extraction of anthocyanins and flavonols from previously lyophilized plant material with a mixture of methanol containing HCl employing ultrasound has been carried out [62]. These natural colorants are useful as food dyes and have application in pharmaceuticals and cosmetics. The mixture is exposed to ultrasound until the extraction is complete and flavonols and anthocyanins are obtained from a column chromatographed extract with a stationary phase of deactivated SiO$_2$-cellulose.

It has been reported that the yield of saponins from *Panax ginseng* is approximately 30% larger after ultrasonic irradiation and the yield of total extract is

Table 11. Ultrasonic Extraction of Chlorophyll

Sample	Time (s)	US Intensity (10^4 W m^{-2})	% Chlorophyll Extracted
Urtica dioica	300	1.5	3.48
		2.0	12.70
		3.5	18.00
	120	1.5	1.39
		2.0	3.48
		3.5	7.14

Table 12. Effect of Sonication Time on the Extraction of Amino Acids from
Amaranthus

Amino Acid	Sonication Time (min)		
	5	10	15
Histidine	0.86	0.50	—
Arginine	1.10	0.86	0.50
Asparagine	0.62	—	—
Glutamic acid	2.66	1.30	1.23
Glycine	1.10	1.04	0.98
Methionine	2.19	0.33	0.33
Leucine	1.47	0.86	—
Tyrosine	0.86	0.43	—
Phenylalanine	0.86	0.43	—

increased by 15% [63]. The yield increases with increasing acoustic power and the saponins themselves are not affected by sonication.

The origin of the use of natural products as pesticides is lost in antiquity and a large number of such materials, used traditionally, still remain to be chemically investigated and evaluated. Although the majority of pesticides used in modern agriculture are synthetic, plant products still contribute to insecticides and rodenticides. Phytochemicals can also serve as lead compounds from which others can be chemically developed to provide greater toxicity toward a particular pest, a wider spectrum of activity against insects, lowered mammalian toxicity, or a decrease in photodecomposition.

One of the main problems facing modern agriculture is to diminish the negative impact of intensive farming which is driven by economic considerations and permitted by the extensive use of chemicals. The increasingly popular alternative methods of organic farming require that synthetic chemicals (particularly pesticides) are removed or replaced with natural biocides, especially of vegetable origin. The best known of the natural insecticides is pyrethrum (composed of chrysanthemic acid esters) and is one of the main constituents of the pyrethrum plant, a member of the chrysanthemum family. Synthetic pyrethrum is now used extensively but the synthetic product is chemically the same as the natural material. Ultrasound has been successfully employed to obtain pyrethrumlike chemicals from *Chrysanthemum cinnerariefolium* in a shorter time and at a greater yield than by classical methods [64]. Most of the material was obtained in the first 10 min of sonication and a longer sonication time did not significantly increase the yield but rather caused some degradation of the extracted compound.

Over the last 6 years a large number of plants in addition to pyrethrum have been tested for the biocidal activity of their extracts. Among these some species such as hop and marigold have produced compounds that are active against the Colorado beetle [65].

2.4 Nutrients and Food Additives

Extracts for use in the flavoring of alcoholic beverages have been produced using ultrasound. For example, 200 g oak chips in 1 liter of water was subjected to sonication at 34 kHz and 50 W cm^{-2} for 10 min [66]. The water was removed and replaced with 250 cm^3 ethanol and 510 cm^3 water. Sonication was repeated for 60 min, raising the temperature of the aqueous ethanol from 20 °C to 35 °C. The extract was removed and the entire procedure repeated. The ultrasonic treatment produced more aromatic aldehydes and increased the total proportion of aromatic compounds compared with conventional technology.

The effect of ultrasound (19.3 kHz) on the extraction of sugar from sugar beet has been investigated [67]. The effect is dependent on the duration of sonication and temperature. In the presence of ultrasound the temperature of the process could be lowered from 70 °C to 50–60 °C without changing the yield of the dry material. At 60–70 °C the reduction in process time was from 60 min to 45 min. Increasing the temperature reduced the effects of the ultrasound and an optimal temperature of 50 °C was reported.

A range of high-frequency ultrasound, 0.8, 1.9, and 3.5 MHz, was employed to study the extraction of sugar from sugar beet and oil from sunflower seeds [68]. The ultrasonic power was varied between 0.05 and 0.5 W cm^{-2}, which increased the yield but also caused some chemical modification of the sugar and oil, diminishing the applicability of this method.

An improvement in green tea quality has been achieved using ultrasound in the pretreatment of freshly collected tea leaves prior to steaming [69]. Such treatment markedly increases the organoleptic quality of green tea. Experiments on the extraction of tea leaves, commercially important as a starting point for the production of instant tea, were performed using ultrasound and the results of classical and ultrasonic methods discussed [13]. As a source of ultrasound a probe system working at different power settings was employed. The influence of water:tea ratio (by mass), temperature, sonication time as well as the ultrasonic power was investigated. Sonication of leaf tea in water at 60 °C for 10 min increases the solid extraction by 20% compared with a control. This improved to 40% when sonication was stopped and the mixture was raised to 100 °C. Extending the sonication time had little effect on the yield. The use of ultrasound shows great promise in improving extraction efficiency at lower temperatures than those normally used. An important finding in this work was that increasing the ultrasonic power over a "limit," a small but less significant improvement was achieved, suggesting that it is necessary to reach an optimum ultrasonic power for the best ultrasonic yield.

Indirect sonication has been found to improve the aqueous extraction of proteins from soya beans [70].

2.5 Miscellaneous Extractions

Numerous herbal cosmetics, cleansers, and medicines use concentrated extracts from plants. Preparations made with herbs usually require the addition of a base solvent such as water or alcohol. Aqueous preparations, such as infusions and decoctions, do not keep very well, but alcohol-based tinctures have a much longer shelf life. The latter have another advantage in that alcohol is a very good solvent for many constituents in plants [71]. Infusions are usually made by steeping the aerial parts of plants in hot water, whereas decoctions are used for the woodier parts of plants and involve boiling them in water for periods of between 15 and 30 min. Tinctures are made by steeping the herb in alcohol and involve conventional methodology such as maceration or percolation.

The application of ultrasonic waves in making preparations from the bark of *Rhamnus frangula* was investigated [72]. The authors noted that ultrasound increased the extraction of active substances, the increase being directly proportional to the acoustic power, time of exposure, and temperature of the extractive solvent. No decomposition of active substances was observed and neat ethanol proved to be a better solvent than aqueous ethanol.

Ultrasound can be successfully used in making infusions from herbs containing alkaloids, like ergot. Powdered ergot was extracted using 500-kHz ultrasound for 20 min at 8 W cm^{-2} [73]. This decreased the extraction time by a factor of 3 in comparison with the USSR Pharmacopoeia method using the same extraction ratio.

Comparative studies of infusions from herbs of *Euphorbia virgata* and *E. semivillosa* obtained by different methods have been reported [74]. Extraction of these herbs with 40–70% ethanol in conjunction with 500-kHz ultrasound at 15 W cm^{-2} was more effective than maceration or percolation. All extractions contained flavonoids, tannins, and hydroxycinnamic acids.

Chamomile *(Matricaria chamomilla)* flowers ground to different degrees were extracted with 70% ethanol by maceration and it was intensified by ultrasound and elevated temperatures (42–45 °C) [75]. This decreased the extraction time 18- to 36-fold, depending on the degree of grinding of the plant material, compared with the extraction by maceration alone.

Although sonication has been proven to be a powerful method for the separation of constituents from solid plant material, some extracts such as berberine show reduced extraction efficiency as sonication time increases [76]. This is due to some degradation of the berberine by ultrasonic radiation.

3. EXPERIMENTAL PROCEDURES

3.1 Extractive Value

One of the parameters that is used to quantify the extraction potential of a vegetable material is its "extractive value." The standard procedure for the determination of extractive value is described in the Romanian Pharmacopoeia Tenth Edition and involves mixing the plant material with solvent normally in a ratio of 5:100 by weight. The mixture is allowed to stand for 23 h followed by 1 h of mechanical stirring. After this a 20-g aliquot is taken off, placed in a preweighed round-bottom flask, the solvent evaporated and dried at 105 °C for 3 h. From the weight of the dry residue the percentage of the extractive value can be determined. For sonicated extracts the mechanical stirring was replaced with 0.5 and 1 h indirect sonication. An Erlenmeyer flask containing the plant–solvent mixture was immersed in a Langford Sonomatic 33-kHz cleaning bath. The temperature was kept at 23 °C by a cooling coil immersed in the bath. After sonication the mixture was filtered and treated as mentioned above resulting in the ultrasonic extractive value.

3.2 Swelling Index

Another parameter important in extraction is the "swelling index." It is defined by the British Pharmacopoeia as the volume in milliliters occupied by 1 g of the vegetable material, after it has been swollen in an aqueous liquid. The swelling process is started by shaking the plant–solvent mixture for 1 min every 10 min for an hour. In the ultrasonic process the manual shaking is replaced by a similar sonication time (total sonication time: 6 min). The volume of the mixtures obtained by both procedures was measured and compared, resulting in the ultrasonic swelling index. The swelling index is an important characteristic of plant material. The higher the index, the higher is the extractive potential of the solvent employed. Increasing this index by means of ultrasound indicates that there will be a positive benefit from using ultrasonic irradiation for the extraction of plant material.

Some of the results obtained giving the extractive value and swelling index, using water as solvent [77], are presented in Tables 13 and 14.

It is clear that even for mint, known as a difficult material for extraction, the extractive value is increased by sonication, demonstrating the effectiveness of ultrasound for the enhancement of extraction.

It can be seen that using ultrasound the swelling of material is improved by 14, 20, and 27%, respectively. This means that the penetration of water inside the cells is much better when sonication is applied.

3.3 Laboratory Equipment

The most used device to perform ultrasonic experiments is without doubt the ultrasonic cleaning bath. For extractions almost any type of cleaning bath can be employed, except in the case of very volatile and flammable solvents. With such a device direct or indirect sonication may be used to obtain extracts. A general setup for a cleaning bath in an indirect type of extraction is presented in Figure 9 [15].

For direct sonication extraction the solvent and plant material are mixed together in the bath tank and ultrasonically irradiated through the bath tank walls. This

Table 13. Extractive Value of Some Plants (Solvent Water)

Method Used[a]	Plant Material Employed		
	Hop Cones	Pot Marigold	Mint
1	24.0	30.0	26.4
2	28.3	30.5	26.5
3	29.5	31.4	26.8

Notes: [a]1 = Classical method 1 h mechanical stirring.
2 = 1/2 h sonication instead of mechanical stirring.
3 = 1 h sonication instead of mechanical stirring.

arrangement is suitable for water or alcohol–water extracts. However, an additional stirring system is often required for a better extraction (see Figure 10) [14–16].

When using a cleaning bath a number of precautions are necessary. First, plant material agglomerates in regions within the bath so that only the outside of the clump of material is exposed to the ultrasonic waves. Sometimes the material sinks, or in the case of a soft plant it will float on the surface of the solvent. These clumps will absorb the ultrasonic energy and this can result in local heating. These effects can be counteracted by the use of efficient mechanical stirring. However, the speed of the stirrer has to be correctly set in order to provide efficient mixing while at the same time avoiding any decoupling effects of the ultrasonic transducers from the rapidly stirred solvent.

Sometimes it is necessary to perform an extraction at a controlled temperature and for this purpose an immersed cooling coil can be used. Since the action of ultrasound produces corrosion, the material of the cooling coil should not be fabricated of copper or aluminum.

The best system with which to perform an ultrasonically assisted extraction is a closed reactor fitted with an ultrasonic probe and mechanical stirring (Figure 11). Usually such a reactor has a device that can control the extraction temperature.

One of the most important factors in such types of experimental arrangements is agitation. Since the ultrasound coming from the horn tip is directional, only a small

Table 14. Swelling Index in mL/L g Plant Material

Method Used[a]	Plant Material Employed		
	Inula helenium	Pot Marigold	Lime
1	7	22	25
2	8	28	30

Notes: [a]1 = Classical method, mechanical stirring.
2 = Sonication instead of mechanical stirring.

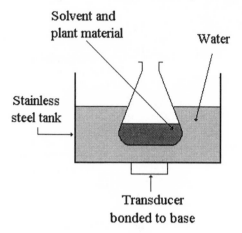

Figure 9. Experimental setup for indirect extraction using cleaning bath.

part of the plant material is in direct contact with the ultrasonic field without agitation. Moreover, the heat that is normally produced by cavitational collapse cannot be removed without agitation, due to the low thermal conductivity of the plant–solvent mixture.

3.4 Large-Scale Ultrasonic Extraction

Large-scale extraction by means of ultrasound is possible in many ways. One of these is to use a large cleaning bath to which mechanical agitation and a cooling

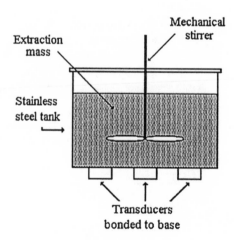

Figure 10. Experimental setup for direct extraction using cleaning bath.

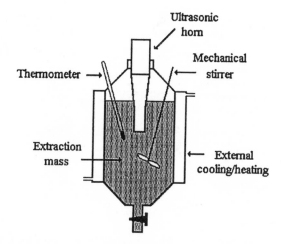

Figure 11. Experimental setup for direct extraction using an ultrasonic horn.

system are attached. It is also possible to use an existing reactor with ultrasonic transducers bonded to the reactor's walls (Figure 12).

Another way is to convert an existing reactor for ultrasonic use with a submersible ultrasonic transducer immersed in the extraction mixture. Several types of submersible transducers are commercially available, to process vegetable material depending on the type of application.

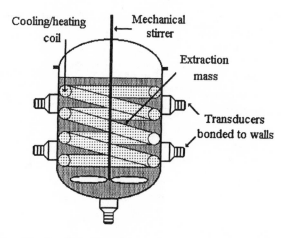

Figure 12. Possible setup for an ultrasonic extraction reactor.

4. GROWING CROPS FOR SPECIFIC EXTRACTS

4.1 A Selection of Important Crops Grown for Commercial Extraction

Some of the most important plants currently commercially grown for producing extracts or essential oils are listed below [78].

- *Anethum graveolens* (dill)

  Plant parts used: Leaf and seeds used in cookery; fruits for essential oil

  Action of extracts: Carminative and stomachic

- *Angelica archangelica* (angelica)

  Plant parts used: Leaves for infusion, tincture, cream; roots for tincture, compress, massage oil; seeds for the same purpose

  Action of extracts: Carminative, antispasmodic, promotes sweating, topical anti-inflammatory, expectorant, diuretic, digestive, tonic, antirheumatic, uterine stimulant

- *Calendula officinalis* (pot marigold)

  Plant parts used: Flowers for infusion and tinctures

  Action of extracts: Antiseptic, anti-inflammatory

- *Carum carvi* (caraway)

  Plant parts used: Fruits for essential oil, flavoring, and spice

  Action of extracts: Antispasmodic, carminative, expectorant

- *Coriandrum sativum* (coriander)

  Plant parts used: Fruits for essential oil, infusion, and spice

  Action of extracts: Aromatic, carminative, antispasmodic

- *Foeniculum vulgare* (fennel)

  Plant parts used: Whole plant for infusion, decoction, tincture, flavoring, and spice

  Action of extracts: Stomachic, carminative, anti-inflammatory

- *Humulus lupulus* (hop)

  Plant parts used: Cones for bitter and flavoring taste of beer, infusion, tincture, compress

  Action of extracts: Sedative, anaphrodisiac, restoring tonic for nervous system, bitter digestive, diuretic, cosmetic

- *Hyssopus officinalis* (hyssop)

Plant parts used:	Aerial parts for infusion, tincture, syrup, and flavoring
Action of extracts:	Expectorant, carminative, relaxes peripheral blood vessels, promotes sweating, reduces phlegm, topical anti-inflammatory, antiviral, antispasmodic

- *Lavandula angustifolia* (lavender)

Plant parts used:	Flowers for infusion, tincture, cosmetics, essential oil
Action of extracts:	Relaxant, antispasmodic, circulatory stimulant, tonic for nervous system, antibacterial, analgesic, carminative, promotes bile flow, antiseptic, insect repellent

- *Levisticum officinale* (lovage)

Plant parts used:	Root and rhizome for liquid extract, cookery
Action of extracts:	Sedative, carminative, diaphoretic, expectorant, antimicrobial

- *Melissa officinalis* (melissa or balm)

Plant parts used:	Aerial parts for infusion, tincture, essential oils
Action of extracts:	Sedative, antidepressant, digestive, relaxes peripheral blood vessels, carminative, antispasmodic

- *Mentha piperita* (peppermint)

Plant parts used:	Aerial parts for essential oil, tincture, infusion
Action of extracts:	Antispasmodic, digestive, tonic, carminative, analgesic, antiseptic, diaphoretic

- *Ocimum basilicum* (basil)

Plant parts used:	Aerial parts for infusion, tincture, syrup, cookery
Action of extracts:	Aromatic, carminative, vermifuge, antibacterial, analgesic

- *Thymus vulgaris* (thyme)

Plant parts used:	Aerial parts for infusion, tincture, syrup, massage oil, essential oil
Action of extracts:	Carminative, antiseptic, antitussive, expectorant

- *Valeriana officinalis* (valerian)

Plant parts used:	Root for maceration, infusion, tincture, compress
Action of extracts:	Sedative, hypnotic, nervine, hypotensive, diuretic

4.2 Plants as Source of Raw Chemical Materials

It is well known that plants are a source of raw material for some industrial applications. One of the best known examples is the rubber tree. In addition, oil plants like sunflower, rape, and castor can be not only a source of food material but also a bulk source of chemicals for the cosmetic and chemical industries.

Some of the plants listed above are also a source of chemical compounds, like linalool from coriander, limonene and carvone from dill seeds, and anethole from fennel seeds. α-Pinene can be separated from turpentine oil, which is extracted from coniferous trees in quite large amounts.

Recently from eucalyptus leaves new compounds have been separated (see structure) and their properties investigated [79], showing antibacterial behavior, especially against cariogenic bacteria. Many more chemical compounds are to be found in plants and their biological potential remains to be investigated.

$R = CH_2$

$R = \alpha\text{-OH}, \beta\text{-Me}$

5. CONCLUSIONS

Many conclusions can be drawn from the information presented in this review. One of the most important is that ultrasound is definitively a powerful tool for extraction technology. What remains to be done is to find satisfactory conditions to fulfill the requirements for a good yield under safe, nondestructive, and easy to monitor extraction conditions.

Ultrasonically assisted extraction can be performed on a large scale as well as on a laboratory scale. Laboratory-scale experiments are very easy to set up, but large and industrial-scale experiments must be carefully designed and conducted. It is clear that carefully selected plants can be grown for their high-value chemical components. Extraction and purification processes could both be improved with ultrasound.

ACKNOWLEDGMENTS

The authors thank the EC for financial support (Copernicus research program ERB-CIPA-CT94-0227-1995) for the research work included in this chapter. M.V. and M.T. thank the Ministry of Research and Technology (Grant 921/396 - 1997) for partial financial support.

REFERENCES

[1] Paun, E., Mihalea, A., Dumitrescu, A., Verzea, M., and Cosocariu, O. *A Treatise of Cultivated Herbs and Aromatic Plants*, Vols. I and II. Romanian Academy Publishing House, Bucharest, 1986; Ionescu-Stoian, P., Stanciu, N., and Savopol, E. *Pharmaceutical Vegetal Extracts.* Medical Publishing House, Bucharest, 1995.

[2] Lewis, D. A. *Chem. Br,.* 28 (1992) 141.

[3] Bojor, O. *Natural Therapy.* Ulpia Traiana, Publishing House, Bucharest, 1995.

[4] Grainge, M., and Ahmed, S. *Handbook of Plants with Pest-Control Properties.* Wiley, New York, 1987.

[5] Guenther, E. *The Essential Oils.* Van Nostrand, Princeton, N.J., 1950, Vol. I, pp. 87–213.

[6] Vitzthum, O. Canadian Patent Can. 987,250 (1976); Chem. Abstr., 85 (1976) 44923; Laws, D. R. *J. J. Inst. Brew.*, 83 (1977) 39.

[7] Information presented at "The potential uses of crop derived raw materials for industry," The National Herb Centre, Banbury, U.K., 1997.

[8] Rapeanu, D. Doctoral Thesis, Bucharest, Romania, 1973.

[9] Hansel, V., and Syre, K. *Pharm. Ztg.*, 106 (1961) 911.

[10] Abramov, O. V. *High-Intensity Ultrasound: Theory and Industrial Applications.* Gordon and Breach, London, 1998.

[11] Ueda, S., Watanabe, A., and Tsuji, F. *J. Electrochem. Soc. Jpn.*, 19 (1951) 142; *Chem. Abstr.,* 45 (1951) 8923a; *J. Electrochem. Soc. Jpn.*, 19 (1951) 193; *Chem. Abstr.,* 46 (1952) 1367c.

[12] Schmall, A. *Schweiz. Brau. Rundsch.*, (1953) 1; *Chem. Abstr.*, 47 (1954) 2932.

[13] Mason, T. J., and Zhao, Y. *Ultrasonics*, 32 (1994) 375.

[14] Vinatoru, M., Stilpeanu, D., Velea, S., Petcu, M., Vija, M., Bacneanu, G., and Brinzan, C. *Tecnol. Chim.*, 1 (1996) 76; *Chem. Abstr.*, 124 (1996) 315365f.

[15] Vinatoru, M., Toma, M., Radu, O., Filip, P. I., Lazurca, D., and Mason, T. J. *Ultrasonics Sonochem.*, 4 (1997) 135.

[16] Salisova, M., Toma, S., and Mason, T. J. *Ultrasonics Sonochem.*, 4 (1997) 131.

[17] Keiki Kamibayashi. Japan Patent 1175 (1950); *Chem. Abstr.*, 46 (1952) 7796.

[18] Ivanova, E. I., and Leontevskii, K. E. *Tr. Vses. Nauchno-Issled. Inst. Zhirn.* 19 (1959) 260; *Chem. Abstr.* 57 (1962) 8675.

[19] Thompson, D., and Sutherland, D. G. *Ind. Eng. Chem.*, 47 (1955) 1167.

[20] Vicci, O. Italian Patent 567, 1957, 201; *Chem. Abstr.*, 54 (1960) 2052.

[21] Rolf, P. German (East) Patent 20, 1960, 312; *Chem. Abstr.*, 55 (1961) 26483.

[22] Schwing, W. F., and Sole, P. *J. Am. Oil. Chem. Soc.*, 44 (1967) 585.

[23] Peichev-Totev, S., and Dimitrova-Tsamena, L. *Mezhdunar. Kongr. Efirnym. Maslam*, 4th, 1968 (pub. 1971), pp. 270–272; *Chem. Abstr.*, 78 (1973) 163929.

[24] Velichkov, V., and Georgiev, E. *Nauchni Tr. Vissh. Inst. Khranit. Vkusova Prom. Plovdiv*, 16 (1969) 1778; *Chem. Abstr.*, 78 (1973) 20103.

[25] Baikoff, E. U.K. Patent Appl. G. B. 2,097,014 (1982); *Chem. Abstr.*, 98 (1983) 18452.

[26] Kachakhidze, G. G. *Tr. Gruz. Politekh. Inst.*, 4 (1970) 44; *Chem. Abstr.*, 76 (1971) 89933.

[27] Goltz, K., Yung, R., and Troger, B. Ger 277, 913 (Cl.C11B9/02) (1990); *Chem. Abstr.*, 114 (1991) 149916.

[28] Specht, W. Z. *Lebens. Unters. Forsch.*, 94 (1952) 157; *Chem. Abstr.*, 46 (1952) 6320.

[29] Schmall, A. *Schweiz. Brau. Rundsch.*, (1953) 1; *Chem. Abstr.*, 47 (1954) 2932.

[30] Arentoft, H., and Arentoft, H. *Brauerei*, 8 (1954) 23/24; *Chem. Abstr.*, 49 (1955) 9873.

[31] Schild, E., and Weyh, W. *Brauwissenschaft*, 13 (1960) 206; *Chem. Abstr.*, 54 (1960) 23176.

[32] Lunenburger Kronenbrauerei. British Patent 788357 (1958); *Chem. Abstr.*, 52 (1958) 8454g.

[33] Hoggan, J. *Ultrasonics*, 6 (1968) 217.

[34] Leoninck, V. *Fermentn. Spirt. Prom.*, 34 (1968) 16; *Chem. Abstr.*, 70 (1969) 18869.

[35] Obolensky, G. *Qual. Plant. Mater. Veg.* 5 (1958) 45; *Chem. Abstr.*, 53 (1959) 11761.

[36] Schultz, O. E., and Klotz, J. *Arzneim. Forsch.*, 4 (1954) 325; *Chem. Abstr.*, 48 (1954) 10301i.

[37] Head, W. F., Jr., Beal, H. M., and Lauter, W. M. *J. Pharm. Sci.*, 45 (1956) 239.

[38] Adamski, R., and Mizgalski, W. *Acta Pol. Pharm.*, 14 (1957) 119; *Chem. Abstr.*, 51 (1957) 14203a.

[39] Ovadia, M. E., and Skauen, D. M. *J. Pharm. Sci.*, 54 (1965) 1013.

[40] Shinyans'kii, L. A., Kazarnovskii, L. A., Karavai, N. Y., and Solon'ko, V. N. *Farm. Zh.*, 3 (1960) 48; *Chem. Abstr.*, 61 (1964) 10533.

[41] Wray, P., and Small, L. D., *J. Pharm. Sci.*, 47 (1958) 832.

[42] Colian, B., and Tomas, G. *Acta Pharm. Hung.* 26 (1956) 67.

[43] Shinyans'kii, L. A., Kazarnovskii, L. A., Karavai, N. Y., and Solon'ko, V. N. *Farm. Zh.*, 2 (1959) 27; *Chem. Abstr.*, 56 (1962) 8838.

[44] Szentessy, S. I. *Gyogyszereszet*, 14 (1970) 133; *Chem. Abstr.*, 73 (1970) 80431.

[45] Zikova, N. Y. Kazarnovski, L. A., Solon'ko, V. N., and Shinyans'kii, L. A. *Farm. Zh.*, 4 (1961) 15; *Chem. Abstr.*, 59 (1963) 3721.

[46] Aimukhamedova, G. B., and Korneva, G. B. *Izv. Akad. Nauk Kirg. SSR Ser. Estest. Tekh. Nauk*, 4 (1962) 17; *Chem. Abstr.* 59 (1963) 15118.

[47] Povikov, A. A., Aimukhamedova, G. B., Kukavishnikova, E. P., Nefedov, A. G., Toktobaeva, F. M., Abdrazakova, N. V., and Mokeeva, B. B. In Afanas'ev, V. A. (ed.). *Razvit. Khim. Proizvod. Kirg.* Ilim, Frunze, USSR, 1976, pp. 102–103; *Chem. Abstr.*, 89 (1968) 48826.

[48] De Maggio, A. E., and Lolt, J. A. *J. Pharm. Sci.*, 53 (1964) 945.

[49] Boiko, V., and Mizinenko, I. *Khim. Farm. Zh.*, 2 (1968) 60; *Chem. Abstr.*, 69 (1968) 5164.

[50] Bose, P. C., and Sen, T. C. *Indian J. Pharm.*, 23 (1961) 222; *Chem. Abstr.*, 56 (1962) 1525.

[51] Srinivasulu, C., and Mahapatra, S. *Indian J. Pharm.*, 32 (1970) 90.

[52] Srinivasulu, C., and Mahapatra, S. *Ultrason. Int. Conf. Proc.*, (1973) 25.

[53] Guo, X., Wang, R., Yun, W., Lin, S., and Hoa, Q. *Shaqanxi Shifan Daxue Xuebao, Ziran Kexueban*, 25 (1997) 47; *Chem. Abstr.*, 126 (1997) 3209539.

[54] Makarenko, P. N. *Vestn. Khar'k. Politekh. Inst.*, 34 (1968) 64; *Chem. Abstr.*, 74 (1972) 34569.

[55] Boeva, A., and Dryanovska, N. *Farmatsiya*, 21 (1971) 33; *Chem. Abstr.*, 76 (1972) 117455.

[56] Morrison, J. C., and Woodford, R. *J. Pharm. Pharmacol.*, Suppl. 19 (1967) 1S.

[57] Patel, I., and Skauen, D. M. *J. Pharm. Sci.*, 58 (1969) 1135.

[58] Rothbaecher, H., and Rothbaecher, E. *Farmacia*, 18 (1970) 39.

[59] Seth, R. K., and Sarin, J. P. *Indian J. Pharm.*, 32 (1970) 11; *Chem. Abstr.*, 70 (1968) 114897.

[60] Naidenova, E., Dryanovska-Noninska, L., and Dimitrov, D. *Probl. Farm.*, 3 (1975) 23; *Chem. Abstr.*, 86 (1977) 185298.

[61] Achimescu, V., Georgescu, C., and Ciobanu, E. *Farmacia*, 30 (1982) 159; *Chem. Abstr.*, 98 (1983) 49985.

[62] Diez de Bethencourt, C., Garrido Valencia, J. L., Reviela Garcia, E., and Santa Maria, B. G. Spanish Patent E.S. 541,893 (1985); *Chem. Abstr.*, 105 (1986) 174396.

[63] Ide, M., and Li, M. *Jpn. J. Appl. Phys. Part 1*, 33 (1994) 30085; *Chem. Abstr.*, 121 (1994) 65390.

[64] Romdhane, M. Ph.D. Thesis, Institut National Polytechnique de Toulouse, 1993.

[65] Wyrostkiewicz, K. Ph.D. Thesis, *Rolnica Rozpravy*, (1992) 53.

[66] Huebner, G., Seliger, E., Petermann, H., and Muelle, M. Offen. 2,836,676 (1979); *Chem. Abstr.*, 90 (1979) 166575.

[67] Gilyus, I. P. *Akust. Ul'trazvuk. Tekh.*, 3 (1968) 40; *Chem. Abstr.*, 71 (1969) 126265.

[68] Ioan, L., Chiril, P., Dumitru, P., and Ionela, P. *Chiem. Ser. I*, 8 (1954) 43.

[69] Okazaki, T. Jpn. Kokai Tokkyo Koho Japanese Patent 60 41 44 [85 41 443] (1985); *Chem. Abstr.* 103 (1985) 52969.

[70] Fukose, H., Ohdaira, E., Masuzova, N., and Ide, M. *Jpn. J. Appl. Phys. Part 1*, (1994) 33.

[71] Mabey, R., McIntyre, M., Michael, P., Duff, G., and Stevens, G. *The Complete, New Herbal.* Penguin Books, 1991.

[72] Kubiak, Z. *Diss. Pharm.*, 14 (1962) 229; *Chem. Abstr.*, 57 (1962) 16745.

[73] Pivnenko, G. P., and Sotnicova, O. M. *Farm. Zh.*, 20 (1965) 39; *Chem. Abstr.*, 64 (1996) 5403.

[74] Soboleva, V. A., Chagovets, K. K., and Solon'ko, V. N. *Farm. Zh.*, 2 (1978) 89; *Chem. Abstr.*, 89 (1978) 185966.

[75] Peric, B., and Lepoyevic, Z. *Zb. Rad-Tech. Fac. Novom Sadu*, 21 (1990) 85; *Chem. Abstr.*, 117 (1992) 118304.

[76] Yuan, M., and Chem, C. *Xiangtan Daxue Ziran Kexue Xuebar*, 15 (1993) 73; *Chem. Abstr.*, 121 (1994) 200180.

[77] Vinatoru, M., and Toma, M. Unpublished results.

[78] Lazurca, D. Plafar Brasov Romania, personal communication.

[79] Osawa, K., Yasuda, H., Morito, H., Takeya, H., and Itokawa, H. *J. Nat. Prod.*, 59 (1996) 823.

POWER ULTRASOUND IN LEATHER TECHNOLOGY

Ji-Feng Ding, Jian-Ping Xie, and
Geoffrey E. Attenburrow

OUTLINE

Advances in Sonochemistry
Volume 5, pages 249–278.
Copyright © 1999 by JAI Press Inc.
All rights of reproduction in any form reserved.
ISBN: 0-7623-0331-X

1. A BRIEF INTRODUCTION TO LEATHER MANUFACTURING

The conversion of putrescible animal hides and skins into stable and useful artifacts known as leather may be man's oldest technology, and the profession of leather technologist or tanner is thus one of the most ancient. Although the industry has a long history, the pace of change has been rapid in this century and has accelerated further in the past few decades. The time required to process raw hide or skin to finished leather has decreased from over a year in the last century to a matter of days now.

The hides (from big animals) and skins (from small animals) used for leather production are mainly the by-products of the meat industry. They are usually preserved with salt. The structure of skin is illustrated in Figure 1. Starting from the hair side there are three distinct layers: grain, corium, and flesh. Of most importance to leather making is the structural protein in skin, i.e., collagen. Noncollagenous components in the skin, such as hair, epidermis, fats, flesh, and interfibrillary proteins and proteoglycans (e.g., albumins, globulins, dermatan sulfate), need to be removed during the early stages of leather making.

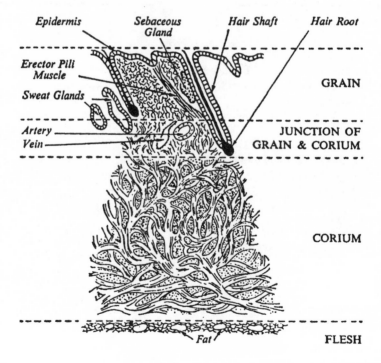

Figure 1. A schematic representation of skin's cross section by Sharphouse [1].

As shown schematically in Figure 1, the skin consists of a network of collagen fibers, very intimately woven and joined together. At the largest level of the hierarchical structure is the fiber bundle (cross-section diameter 60–200 μm) [2]. The fibers (30–60 μm) are made of fibril bundles (3–6 μm) with fibrils (100–200 nm) being the finest level of structure that is visible by scanning electron microscopy (SEM) in intact collagen. Figure 2 shows a SEM micrograph of a fiber, composed of fibril bundles [3]. In the grain layer these fibers are thin and tightly woven, and so interlaced that there are no loose ends on the surface beneath the epidermis. Thus, when the epidermis is carefully removed, a smooth surface is revealed, which gives the characteristic grain surface of leather. Toward the center of the corium layer the fibers are coarser and stronger, and the predominant angle at which they are woven can indicate the properties of the resultant leather. If the fibers are more upright and tightly woven, one may expect a firm and hard leather, whereas if the fibers are more horizontal and loosely woven, one may expect a soft leather. The fibers on the flesh side of the corium have a more horizontal angle of weave.

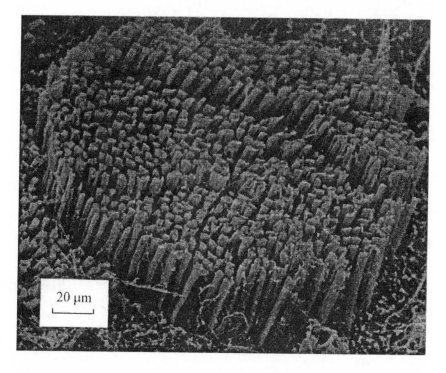

Figure 2. A fiber, composed of fibril bundles, emerging from the ice surface in a cryo-SEM photomicrograph (reproduced with permission [3]).

In addition to various mechanical operations such as fleshing, drying, and staking, leather is made from hides and skins by about a dozen major processes carried out in an aqueous medium (so-called wet processes) using wooden drums as the vessels. These processes are briefly introduced below.

- *Soaking*.   This first process is to wash and completely rehydrate salted hides or skins.
- *Unhairing* and *liming*.   These two processes may be carried out at the same time or separately to remove hairs, epidermis, and the interfibrillary components (especially dermatan sulfate) and to open up the fiber structure. The hairs on the skins or hides are chemically removed, usually by sodium sulfide buffered with calcium hydroxide. Then the structure is opened up by prolonging the treatment in saturated lime or calcium hydroxide at pH 12.5 (liming). Liming also converts the amide group of asparagine and glutamine to a carboxylate group, thus increasing the number of carboxylic acid groups available for tanning. As a result, limed pelts have a lower isoelectric point (pH 5–6) than that of native skins (pH ~9).
- *Deliming*.   This is to neutralize and remove the alkali used in the liming process. The pH often needs to be lowered to between 8 and 9 for the bating process.
- *Bating*.   This is a treatment of the hides or skins with proteolytic enzymes to break down the noncollagenous proteins albumins and globulins.
- *Degreasing*.   This process is to remove the fats in the skins, preferably in an aqueous medium by using either surfactants or enzymes.
- *Pickling*.   This is to bring the skin to the right acidity for tannage with sulfuric acid in brine.
- *Tanning*.   Tanning is the process by which putrescible skins are converted into stable materials resistant to microbial attack and wet and dry heat. Tanning is regarded as having been accomplished when an expected final shrinkage temperature (T_s) is achieved. The shrinkage temperature is the temperature at which a specimen shrinks on heating when it is held in water. There are many different tannages and the choice depends chiefly on the properties required in the finished leather.

    Current tanning technology is dominated by chromium(III) or chrome tannage. Chrome tanning is usually initiated at pH 2.5–3.0, using basic chromium(III) sulfate, which penetrates through the skin. The pH is then raised to 3.5–4.0 (basification). Increasing the pH increases the reactivity of both collagen and chromium, leading to a reaction between the carboxy groups on collagen and polymerized chromium species. After chrome tanning the leather is called *wet-blue*, because it is wet and has a characteristic blue color due to the presence of Cr(III).

    The traditional tannage is vegetable tanning based on plant polyphenols (so-called vegetable tannins). It is still widely used. Effective vegetable tannins have a molecular weight of 500–3000; they interact with collagen primarily via hydrogen bonding [4]. The penetration of these large molecules of high affinity to proteins is slow and the whole process may take a few weeks.

Other tannages include aluminum, aldehyde, oil (polyunsaturated oil), syntan (synthetic tanning agent), and organic tannage (using synthetic polymers and cross-linkers).

- *Retanning.* Often a single type of tannage such as chrome tanning alone may not be sufficient to provide the required properties of finished leather. So a secondary tannage is needed after the prime tannage. A chrome tanning may therefore be followed by a retanning with vegetable tannins or syntan.
- *Neutralizing.* This is to adjust the interior pH of leather after tanning as a necessary preparation for dyeing or fatliquoring.
- *Dyeing.* Acid dyes are the most commonly used dyestuff in leather dyeing. They are anionic, thus the surface charge of leather will influence the dyeing process. The surface charge of leather can be controlled by pH. Usually, dyeing is initially carried out at a pH above the isoelectric point, which varies depending on the tannage(s) used. This results in a negative surface charge for the leather so as to facilitate the dye penetration. The pH is then lowered to assist the fixation of dye to leather.
- *Fatliquoring.* This is the process by which oil is applied through an oil-in-water emulsion to lubricate the leather fibers so as to achieve the required softness and "handle." Emulsifiers used are mainly anionic, thus pH control is important for the process.

Clearly all of these processes involve mass transportation into or out of a woven fiber network through diffusion. It is conceivable that power ultrasound can enhance all of these processes.

2. EARLY INVESTIGATIONS ON THE EFFECTS OF ULTRASOUND ON LEATHER PROCESSING

The potential benefit of power ultrasound to leather processing was recognized shortly after World War II. The earliest patent in this field was by Zapf in 1949 [5], which first claimed the use of ultrasound for soaking, liming, bating, tanning, dyeing, and oiling. Most of the early work was carried out in the 1950s and 1960s, and the number of early publications [5–25] in this field is limited. At that time, ultrasonic equipment was not widely available so self-constructed ultrasonic apparatus and a wide range of frequencies were employed in many investigations. Most investigations [5–14, 19–25] were concerned with tanning, soaking, unhairing and liming, fatliquoring, and dyeing, although degreasing [12], bating [5,15] and oiling [5,10] were also examined. In this section some of the early work (up to 1978) is reviewed.

2.1 Soaking, Liming, and Unhairing

Zapf [5] was one of the pioneers who explored the use of ultrasound in leather processes. In his patented processes, ultrasound (50–3000 kHz) was applied from above and below to horizontally suspended moving hides. Irradiation at intervals and with variable frequency was recommended. For soaking, he suggested an ultrasonic power of 8 W cm^{-2} and irradiation periods of 3 × 3 min at 22 °C.

In the early 1950s, Fridman et al. [8] applied ultrasound to the liming process. In their experiments, a transducer with a frequency of 1200 kHz and a power of 8–10 W cm^{-2} was employed in a reaction vessel equipped with a cooling system. They found that, after ultrasonic treatment for 6 h at 30 °C, the hair and epidermis came off easily and the grain surface was smooth, whereas hair and epidermis were not removed in this time by a lime suspension without ultrasound.

In 1956, Realisations Ultrasoniques [9], a French company, patented a technique that claimed that ultrasound can be successfully used to remove hairs. A dry skin was first soaked for at least 50 h in a water bath containing a wetting and an antiseptic agent. The skin was then dried slightly before being dipped into a water tank and treated by ultrasound (10–100 kHz). A 3-min treatment was usually sufficient to cause the ligaments holding the hair bulb to become effectively separated. Thereafter, the removal of hairs from this ultrasound-pretreated skin was reportedly more readily effected using normal procedures. It was also claimed that the hair and skin did not suffer from any damage or degradation and that the treated skins were very easy to tan.

At the same time, Mieczyslaw [10] also studied the effect of ultrasound (300 kHz to 3 MHz) on soaking, liming, and oiling of calf skins. He found that soaking time was reduced to 1 h and liming time was reduced to 1.5 h, compared with 18–20 h for the standard process. In some cases the hair was loosened during the ultrasonic soaking.

However, when Herfeld [11] reexamined the use of ultrasound in unhairing and liming some 20 years later, his conclusion was quite different from the previous workers. He observed that, although the hair of ultrasonically soaked pelts was considerably cleaner, the increase in water uptake was only 10–15%. For liming, ultrasound produced a slight increase in the diffusion rate of liming chemicals. This would not justify the use of ultrasound in view of the increased cost and the risk of grain damage to the untanned pelts. He also found that there was no difference in the time of hair-loosening for ultrasound of between 22 and 40 kHz.

2.2 Tanning

Tanning is one of the most time-consuming processes in leather making, especially in the case of vegetable tanning for heavy leather, which requires a few weeks. Therefore, the use of ultrasound to speed up the tanning process was attempted as early as 1949 by Zapf [5]. The skins were laid on a conveyor belt in a shallow bath filled with tanning liquor, and subjected to ultrasound irradiation from above or below; alternatively the skins were suspended, and subject to a moving transducer.

In 1950 Ernst and Gutmann [6] also used ultrasound to assist the chrome, vegetable, and synthetic tannages for calf skin. The pelts were suspended horizontally in the liquors, parallel to and about 2.5 cm distant from the surface of the quartz transducer. The transducer was excited from a conventional power oscillator, and a power output of 15.5–50 W was employed. The ultrasound had a frequency of 760 kHz. Retuning of the oscillator was necessary to maintain the frequency as the resonant frequency shifted during the irradiation period. The experiment was carried out at room temperature (± 2 °C). After 3.5 h vegetable tanning with ultrasound treatment, the pelts showed a very considerable speeding up of the tanning process and already exhibited a leatherlike appearance. The tanning agent had deeply penetrated; the depth of penetration was found to vary between 20 and 25% of the thickness. The control samples (no ultrasound), however, showed very little trace of either penetration of the tanning agent or leatherlike appearance. Later in a separate paper, Gutmann [7] revealed that the ultrasonic power was about 4.5 W cm^{-2} for the vegetable tanning experiments. Results for syntan (synthetic tanning agent) were even more positive than for vegetable tannage. However, chrome tanning did not show an appreciable rate increase. But the tanning agent appeared to be much more uniformly deposited on the skin surface, as could be seen from the darker and more even color than that in the control sample.

Ernst and Gutmann [6,7] proposed that the reason for a lack of improvement in chrome tannage was due to the small size of complex chromium ions, even if hydrated. In comparison with the width of channels separating adjacent polypeptide chains in collagen, which is on the order of 15 Å, the radius of the unhydrated chromium ion is only 1.62 Å. It therefore should not be difficult for the chrome complex to pass through these channels. As a result, there is not much scope for ultrasound to speed up the process of penetration. For vegetable tanning the reduction of aggregate size and the depolymerizing action of ultrasound were believed to be responsible for the increased penetration. It was also surmised by them that the large "frictional force" produced at the pelt–liquor interface would tend to break up any absorbable clusters of tanning agent which otherwise could clog the pelt openings and slow down the diffusion process. Also it was felt that the ultrasound would tend to open the residual linkages between opposing side chains of the collagen molecule thereby increasing the number of reaction sites available for the tanning chemicals.

Tielborger [12] also patented the use of ultrasound (370–800 kHz) for shortening the tanning process, including vegetable, chrome, and aldehyde tanning. Use of ultrasound for tanning was also claimed by Eisenegger in a Swiss patent [13]. Fridman et al. [8], using a high-frequency ultrasound (1.2 MHz, 8–10 W cm^{-2}), found that at 30 °C tanning was completed in 18 h with, and 114 h without, ultrasonic action. Mieczyslaw [10] reported that the rate of tannage was greatly accelerated under the influence of ultrasound (200 kHz–3 MHz). For instance, sulfited quebracho and a syntan material ("Tanigan Extra A") completely penetrated a 7-mm-thick calfskin in 8 h with ultrasound, but for an ordinary process it took 12 h to penetrate half the thickness. In an oil tannage the oil was emulsified, oxidized, and then reacted faster under irradiation with ultrasound.

Ultrasonic preparation of emulsions for chrome emulsion tanning was reported by Aksel'band et al. [14]. After the tanning it was found that the Cr_2O_3 content in the goatskin leather was 4.8%, compared with 3.5% for a normal tanning with the same amount of Cr_2O_3 offer.

In addition to ultrasound, it is worthwhile mentioning that the effect of sound waves on leather processes was also examined in the early 1950s. Simoncini and Criscuolo [26] tried to apply a sound wave (100 Hz), well outside the ultrasonic range, in vegetable tanning. The experiment was carried out in a cylindrical vessel (15 liters) of stainless steel, irradiation from the bottom. In the case of chestnut extract tanning, a degree of tannage of 64% was achieved in 96 h (with 23 h of irradiation), compared with 144 h without irradiation. If tanning for 144 h was used with 33 h of irradiation, the degree of tannage increased to 82%. Potoschnig and Liebscher [27] observed a more uniform penetration and a more rapid tannage using sonic vibrations (1–16,000 Hz), and they claimed the same technique was applicable to soaking, liming, and bating processes. A similar investigation on tanning was also carried out by Schantz [28] with a 100-Hz sound wave. Schantz argued that, compared with ultrasound, sound has no risk of damaging skins and is more suitable for scaleup.

Quite interestingly, Mieczyslaw [10], in investigating the use of ultrasound in vegetable tanning, found that quebracho extract subjected to ultrasound showed a 7% rise in tannin content, which was independent of frequency (200 kHz–3 MHz). The maximum effect was at 62.5 °C. The result was interpreted as a depolymerization of phlobaphenes. Witke [15] reported the application of ultrasound to the extraction of tanning material at room temperature and the resulting products had better diffusion properties. Many other authors, Karpman [16], Alexa et al. [17], and Khr [18], also claimed that ultrasound can increase the tannin content, reduce the insoluble fraction and viscosity, or result in a more rapid extraction.

The application of ultrasound in a tannery was attempted by Wenzinger [19] and Masner [20]. Masner reported that vegetable tanning could be shortened to 20–48 h by employing ultrasound using magnetostrictive transducers in tanneries, and it was claimed that chrome tanning could also be considerably shortened. Wenzinger tried to develop an automated line that involved the use of a moving belt to convey the hide from one vat to another. This would only be feasible if the tanning processes could be accelerated considerably. He claimed that this could be accomplished by the use of ultrasound and that a 3-mm-thick hide was completely penetrated by quebracho in 30 min under ultrasonic treatment, while a control hide sample showed practically no penetration after the same period of time. Despite the encouraging results of Masner and Wenzinger, ultrasound was not applied commercially in the tannery, probably due to difficulties in scaling up, the immaturity of the technology, and the high cost of equipment at that time.

2.3 Fatliquoring and Dyeing

The early investigations concerned with fatliquoring were focused on the preparation of fatliquor emulsions. This would be an obvious option, in view of the emulsifying power of ultrasound and the convenience of application. One of the early investigations was by Gourlay [21] who reported that ultrasound could be used to emulsify fatliquor quickly and economically. Senilov and Obukhov [22] and Metelkin and Suchkov [23] also successfully prepared fatliquoring emulsions by ultrasound.

Kotlyarevskaya et al. [24] used an acoustic pipe of the Polman type for the preparation of fatliquor emulsion. The emulsion was prepared at 40–45 °C by a 5- to 15-min circulation through the pipe. The preparation of 50 liters of emulsion at 30–40 °C and 3–5 atm at a distance of 2–4 mm from the resonator nozzle took only 3–5 min. The emulsion was stable for 5–7 days. More than 70% of the particles had a diameter of less than 1.5 μm and very few exceeded 5 μm.

As far as leather dyeing is concerned, surprisingly few publications could be traced. This is in sharp contrast to textile dyeing with ultrasound, which was an area of considerable activity from the 1950s to the 1980s [43]. Zapf [5] patented dyeing of moving leathers on a conveyor belt subjected to ultrasound irradiation (e.g., 4-min irradiation with a power of 8 W cm^{-2} at 60 °C). Later, Tielborger [12] patented in 1954 a dyeing process facilitated by ultrasound of 150, 395, and 962 kHz, but little detail was given.

3. RECENT INVESTIGATIONS ON THE EFFECTS OF ULTRASOUND ON LEATHER PROCESSING

The early work in the 1950s and 1960s demonstrated that power ultrasound can enhance many wet processes in leather manufacturing. But interest faded away due to a combination of factors: immaturity of the ultrasonic technology, lack of commercial availability of ultrasonic equipment, high costs and difficulties in scaleup, and so forth. Its industrial application was considered unjustified at that time. Since the early investigations in the 1950s and 1960s, almost 25 years of silence followed!

However, circumstances have changed. On the one hand, the ultrasonic technology has advanced and the commercial availability of equipment has increased significantly, bringing down the costs. Developments in the theory and practice of ultrasound have improved its reliability and reproducibility. The technology is maturing so that it is within the reach of commercial application. On the other hand, increasingly stringent environmental legislation and a much more competitive market demand new efficient technologies for the leather industry. This has led to a resurgence of interest in the use of power ultrasound for leather processing [29–37]. In addition, the general applicability across a breadth of wet processes in leather making should make the application of ultrasound more viable.

3.1 Liming and Unhairing

To reduce the environmental impact of effluent from the liming and unhairing process, much modern unhairing technology is based on hair saving [38,39] rather than hair dissolving, although the process still uses Na_2S/NaHS and lime. Keratin (hair) contains a disulfide bond (cystine), which can be transformed under an alkaline condition into a much more stable thioether bond (lanthionin). This results in an "immunization" of the hair. Hair-save technology relies on a faster immunization of hair shaft than hair root. When the hair shaft is largely immunized, Na_2S or NaHS is added, which acts to prevent the formation of lanthionin in the hair root. Thus, hair can still be removed but almost intact. This allows easy recovery of hairs from the liquor.

Quite recently, ALPA S.p.A [30] in Italy successfully applied ultrasound to this hair-save unhairing process on a commercial scale, producing 100 hides using a newly developed drum equipped with nickel/iron magnetostrictive transducers. Three arrays of transducers, triangularly arranged, were located inside the drum on the wall, as shown in Figure 3. They only irradiate when immersed in liquor. The process was fully patented. It was claimed that ultrasound enables a rapid penetration of a minimum amount of HS^- directly to the hair root after immunization of the hair shaft with lime. This allowed complete hair removal with ultrasonic treatment for 3 h, followed by normal overnight liming. The low level of residual HS^- ensures that the hair shaft remains more intact. The chemical-oxygen-demand (COD) levels were therefore reduced compared with a conventional hair saving process. The typical COD from a conventional process is between 36,000 and 65,000 mg/L, but the COD from the ultrasound process was considerably reduced to 17,200–23,700 mg/L. At the end of liming, the swelling was uniform and the skins were clean and white. The appearance and softness of the limed skins were more similar to those found for a 36-h liming period than an 18-h process.

The main advantage, or emphasis, of this particular ultrasonic processing, as the author argued, is not about shortening process time, which has been the focus of early investigations, but about reduced environmental impact and thus the costs to the industry.

A group from Romania [37] has explored the use of ultrasound (20 and 360 kHz) as a primary treatment for liming effluent from the tannery. This was to destroy the sulfide or hydrosulfide ions in the effluent, which can generate toxic hydrogen sulfide when pH drops. It was claimed that ultrasonic treatment for a very short time can convert sulfide species to sulfur, due to the action of free radicals from the cavitation. It was reported that the sulfide levels were lowered from 200 mg/L to below 1 mg/L.

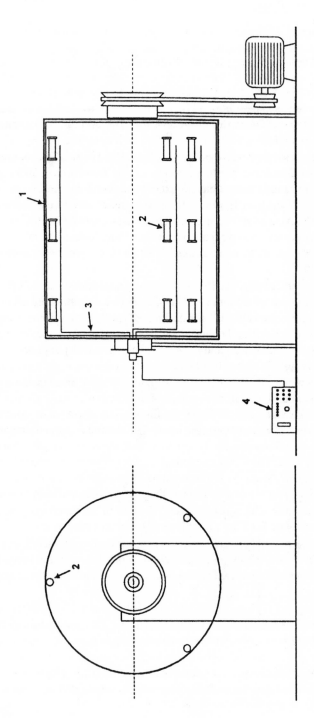

Figure 3. An industrial drum equipped with ultrasonic transducers, developed by ALPA S.p.A [30]. 1, drum; 2, transducer; 3, electrical wire; 4, generator.

259

3.2 Dyeing

In the past 3 years, a research group at the British School of Leather Technology (BSLT) at Nene College, in collaboration with Coventry University, has been actively investigating the effect of power ultrasound on various leather processes including tanning, dyeing, and fatliquoring [29,31,33–36]. The dyeing process was studied in detail, and some of the factors influencing the effectiveness of ultrasonic dyeing were closely examined.

Experiments were carried out on a laboratory scale using the apparatus shown in Figure 4. A side-mounting submersible transducer (KST360, Kerry Ultrasonic Ltd.) was employed. The ultrasound had a frequency of 38 kHz and a maximum power output of 700 W (1.36 W cm^{-2}). The reaction vessel (a cuboid glass container or a small stainless steel drum) was immersed in the temperature-controlled water bath, with its flat surface facing the transducer and about 1 cm away. Detergent (1%) was added to the water bath (not in the reaction vessel) for an effective transmission of ultrasound energy [40].

Different types of leather were dyed with and without ultrasound at different temperatures, using mainly acid dyes. Leather samples were neutralized to pH 6.0–6.5 before dyeing. Typically a 4% offer of dyestuff and 1500% float or water (0.4 g dye, 150 mL water, and 10 g leather) were used in the dyeing experiments. In the leather industry, the amount of dye or chemical used is usually expressed as a percentage of the weight of the leather.

The effect of ultrasound (US) on dyeing was initially examined in a small glass vessel under a static condition (i.e., no mechanical agitation). A very significant increase in both dyeing rate and dye penetration was observed when power ultrasound was applied. In fact, the increase in dye penetration is spectacular, as

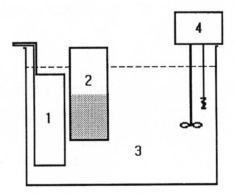

Figure 4. A schematic illustration of the laboratory-scale ultrasonic apparatus used by Xie [35] for leather dyeing, fatliquoring, and tanning. 1, submersible transducer; 2, processing vessel; 3, water bath; 4, temperature control unit.

illustrated in Figure 5. In many cases, the increase in the degree of dye penetration is more profound than that in dye uptake, so as a result the color becomes lighter if ultrasound is used. Figure 6 shows typical plots of dye uptake versus time at two different temperatures (30 and 60 °C). The dye uptake at 30 °C with ultrasound is better than that at 60 °C without ultrasound. If the process is compared at the same

Figure 5. Optical micrographs of dye penetration in bovine wet-blue leather, (top) with ultrasound and (bottom) without ultrasound, at 50 °C for 60 min dyeing with an acid dye (Airedale Brown 3RG).

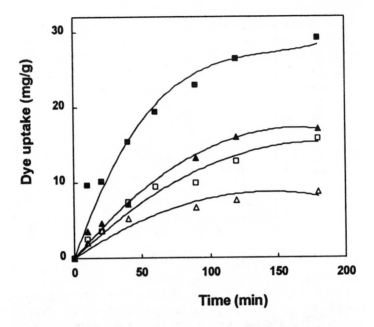

Figure 6. Plots of dye uptake versus time for chrome-tanned and mimosa-retanned sheepskin with an acid dye, Airedale Brown 3RG, (▲) with and (△) without ultrasound at 30 °C and (■) with and (□) without ultrasound at 60 °C.

high temperature (60 °C), the use of ultrasound can shorten the dyeing time by more than 70%.

Leather dyeing is believed to be a diffusion-controlled process. Using a diffusion model in which the substrate is treated as an infinite thick slab of material before the dye penetrates to the center, Vickerstaff [41] developed an approximate equation,

$$C_t = 2C_\infty \, (Dt/\pi)^{1/2}$$

where t is time, C_t and C_∞ are, respectively, the dye uptake at time t and infinity (i.e., dye uptake at equilibrium), and D is the diffusion coefficient. This equation indicates a linear relationship between the dye uptake and the square root of time. As can be seen in Figure 7, replotting of the dye uptake data in Figure 6 versus $t^{1/2}$ shows clearly that this linear relationship holds well indeed for both ultrasonic and nonultrasonic dyeing. Since C_∞ is the equilibrium dye uptake, it is unlikely to be affected by ultrasound. Thus, the effect of ultrasound is to increase the diffusion coefficient for the dye penetration, which consequently increases the dye uptake. This important conclusion is further confirmed by a simple experiment where leather powders are dyed instead of pieces of leather. Due to the large surface area

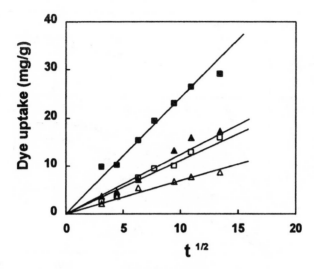

Figure 7. Plots of dye uptake versus square root of time, *t*, for chrome-tanned and mimosa-retanned sheepskin with an acid dye, Airedale Brown 3RG, (▲) with and (△) without ultrasound at 30 °C and (■) with and (□) without ultrasound at 60 °C.

and very small dimensions of these powders, the dyeing will be essentially an adsorption-controlled rather than a diffusion-controlled process. As shown in Figure 8, the effect of ultrasound on dyeing of leather powders is almost nonexistent (cf. Figure 6), indicating that ultrasound, at least at a power of 1.3 W cm^{-2}, has little influence on dye adsorption. Thus, we concluded that ultrasound influences the dyeing process mainly by increasing the diffusion coefficient of the dye penetration process, probably due to the cavitation effect.

If C_∞ is assumed to be independent of whether or not ultrasound is applied, the relative increase of diffusion coefficient can be estimated from the slopes of the plots in Figure 7. For the retanned sheepskin dyed with an acid dye, Airedale Brown 3RG (Yorkshire Chemicals), the ratio of the diffusion coefficient with ultrasound to that without ultrasound was about 3 at 30 °C. However, when the diffusion coefficient was measured using a dialysis tube by Barrer's method [41], this ratio was found to be about 1.6 at 25 °C and 1.2 at 40 °C in an aqueous solution of this dye [35], much less than 3, the value observed in leather dyeing. Thus, the increase of dye diffusion in leather cannot be satisfactorily explained by an increase of dye diffusion in water. How ultrasound influences the dye diffusion in a woven fiber network such as leather remains an interesting question to be answered.

The influence of ultrasound on the rate of the dyeing process varies from one dyestuff to another and from one type of leather to another. As shown in Figure 9, the ratios of dye uptake with ultrasound to that without ultrasound for the retanned

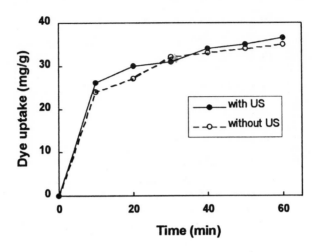

Figure 8. Effect of ultrasound on the dyeing of chrome-tanned leather powders with an acid dye, Airedale Brown 3RG, at 40 °C.

sheepskin at 50 °C are plotted versus time for two acid dyes (Airedale Brown 3RG and Black E793, supplied by Yorkshire Chemicals) and a reactive dye (Procion Blue MX-7RX, supplied by ICI Fine Chemicals). Without ultrasound, the dye uptake for Airedale Black is clearly lower than that for Airedale Brown or Procion Blue, with the dye uptakes for Airedale Brown and Procion Blue being very similar. The relative increases of dye uptake for Airedale Brown and Procion Blue, after ultrasonic irradiation, are very similar too, whereas the relative increase of dye uptake for Airedale Black is distinctly higher than that for Airedale Brown or Procion Blue. This suggests that the effectiveness of ultrasound in enhancing the dyeing process is related to the difficulty of the conventional dyeing process in the absence of ultrasound. The more difficult the conventional dyeing, the more effective is ultrasound in speeding up the process. Similarly, the effectiveness of ultrasound is also influenced by the type of leathers being dyed. When a crust leather (fatliquored bovine wet-blue leather), a chrome-tanned and mimosa-retanned sheepskin leather, and an acrylate resin-retanned bovine wet-blue leather are dyed with the same acid dye without ultrasound, the rank of order in terms of dye uptake is: crust leather > retanned sheepskin > resin-retanned wet-blue. After applying ultrasound, the relative increase of dye uptake is in a reversed order, as shown in Figure 10. The relative increases of dye uptake for retanned sheepskin and resin-retanned wet-blue are very close, and their dye uptakes without ultrasonic irradiation are close to each other as well. This suggests, again, that ultrasound is more useful or effective when a dyeing process has greater difficulty in achieving penetration or diffusion.

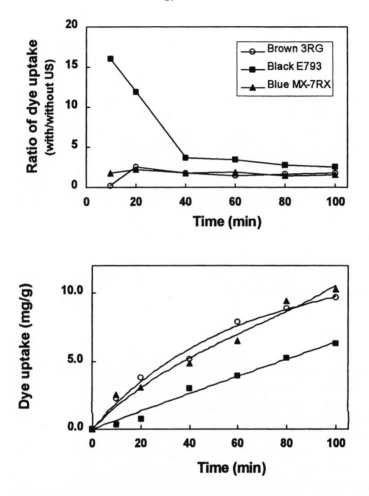

Figure 9. Influence of different dyestuffs on the effectiveness of ultrasonic dyeing of mimosa-retanned sheepskin leather at 50 °C. Top graph shows the ratio of dye uptake with ultrasound to that without; bottom graph shows the dye uptake without ultrasound. Two acid dyes (Airedale Brown 3RG and Airedale Black E793) and one reactive dye (Procion Blue MX-7RX) were used.

Temperature also influences the effectiveness of ultrasound in enhancing the dyeing process. But the effect is more complicated and there seems to be no clear overall trend for all types of leather. Figure 11 illustrates this point. The ratio of dye uptake with ultrasound to that without is again used as an indicator for effectiveness, and plotted versus time for the dyeing of a wet-blue and a retanned sheepskin with the same acid dye at different temperatures. In the case of wet-blue, it is clear that

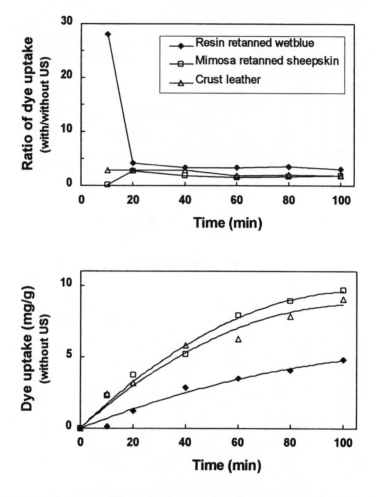

Figure 10. Influence of different leathers on the effectiveness of ultrasonic dyeing with an acid dye, Airedale Brown 3RG, at 50 °C. Top graph shows the ratio of dye uptake with ultrasound to that without; bottom graph shows the dye uptake without ultrasound. An acrylate resin-retanned bovine wet-blue leather, a chrome-tanned and mimosa-retanned sheepskin leather, and a crust leather were used.

the effectiveness of ultrasound as a processing aid decreases as temperature increases. This may be expected as the temperature has a similar effect to ultrasound in increasing diffusion. In the case of retanned sheepskin, a more complicated picture emerges: The order of effectiveness apparently varies with time, but overall it is the highest temperature (60 °C) that gives the highest effectiveness. The reason is unclear, although it is probably related to the retanning reagent (mimosa) used.

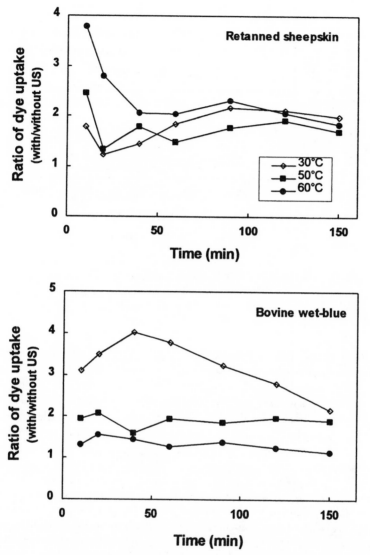

Figure 11. Influence of different temperatures on the effectiveness of ultrasonic dyeing of chrome-tanned and mimosa-retanned sheepskin leather and bovine wet-blue leather.

Interestingly, it was found that initial sonication is much more effective than the later treatment, as shown in Figure 12. In terms of dye uptake, the initial 20-min irradiation produces a result almost similar to a 60-min continuous treatment. Thus, the timing appears to be a very important factor in optimizing the ultrasound effect.

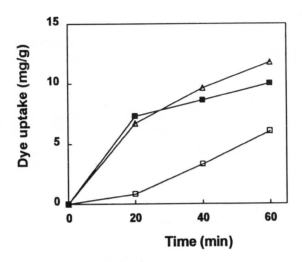

Figure 12. Effect of timing on dyeing kinetics—(△) continuous sonication, (■) first 20 min sonication, and (□) last 20 min sonication—with 4% acid dye (Airedale Brown 3RG) and resin-retanned bovine wet-blue leather at 50 °C.

This finding has considerable practical implication for the energy consumption and the lifetime of ultrasonic equipment.

Another advantage of using ultrasound is that the same level of dye uptake can be achieved with a lower dye offer within the same time scale. A 4% dye offer with ultrasound results in a 40–70% higher dye uptake than that achieved by a 6% offer without ultrasound [35]. This is perhaps more significant than shortening the time, because dyestuff is one of the most expensive chemicals used in the leather industry.

In practice there is no dyeing process that is carried out without any mechanical agitation. The question of how ultrasound compares with mechanical agitation needs to be answered. Figure 13 shows the effect of mechanical stirring (50 rpm) on the dyeing or ultrasonic dyeing. It can be seen that ultrasound is more effective than the mechanical stirring, but a combination of ultrasound and stirring produces the best result. Figure 13 also appears to suggest that the effects of ultrasound and stirring may be additive.

The most effective mechanical agitation for leather dyeing is drumming. A small stainless steel drum was immersed in the water bath, with its flat side facing the transducer. When the drum was used in the case of wet-blue leather and the acid dye, Airedale Brown 3RG, as shown in Figure 14, the effect of ultrasound, apart from the first 10–20 min, was not significant. However, it should be pointed out that when the drum is used, the real irradiation time is only about one-third of the total elapsed time. In addition, it was surprisingly found that the ultrasonic power transmitted into the metal drum was much less than that into a glass vessel (the test

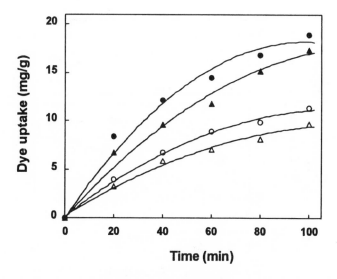

Figure 13. Effect of ultrasound and mechanical stirring on the dyeing of crust leather with acid dye, Airedale Brown 3RG, at 50 °C, (△) without either ultrasound or stirring, (○) with mechanical stirring only, (▲) with ultrasound only, and (●) with both ultrasound and stirring.

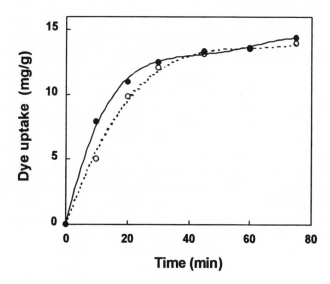

Figure 14. Dye uptake of bovine wet-blue leather with 4% acid dye (Airedale Brown 3RG) and 200% float at 40 °C, using a stainless steel drum, (●) with and (○) without ultrasound.

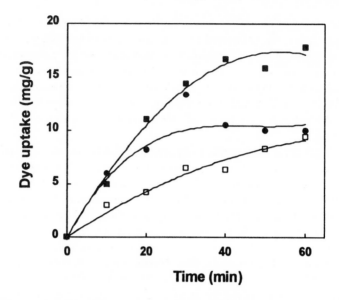

Figure 15. Dye uptake of bovine wet-blue leather with 4% acid dye (Acid Orange 74) and 500% float at pH 6, using a stainless steel drum, (●) at 25 °C with ultrasound, (□) at 60 °C without ultrasound, and (■) at 60 °C with ultrasound.

by aluminum foil showed the number of pits on the foil was 30–50% less than that in a glass vessel). Nevertheless, in the case of wet-blue leather and the acid dye, Acid Orange 74 (supplied by Ciba-Geigy), a significant enhancing effect was observed [36]. As can be seen in Figure 15, the dye uptake at room temperature with ultrasound is even better than that achieved by conventional dyeing at 60 °C. This acid dye is known to be a "difficult dye" that is quite hydrophobic and tends to form aggregates. This reinforces the view made earlier that ultrasound is more effective and useful in dealing with a difficult dyeing system. It has been experimentally shown that the particle size of this dye is reduced by ~40% at 25 °C after applying ultrasound [36].

As far as the quality of the dyed leather is concerned, ultrasound improves the wet and dry rub fastness of the dye, particularly the wet rub fastness. This is probably due to a deeper penetration of dye into the fiber.

3.3 Fatliquoring

As has already been mentioned, the early investigations focused on the utilization of emulsifying power of ultrasound for the preparation of fatliquor emulsions. Xie et al. [31,34,35] recently studied the effect of power ultrasound (38 kHz, 1.3 W cm^{-2}) on the fatliquoring process itself as well as the effect of sonication of fatliquor emulsions on the fatliquoring process. Wet-blue leather (thickness 1.9–2.3 mm) from bovine hides was fatliquored using an anionic fatliquor reagent, Remsynol ESI (Hodgson Chemicals). Samples were neutralized before fatliquoring. The fatliquoring process was carried out in a glass container, with an 8% offer of Remsynol ESI based on the wet weight of leather (typically 20 g leather, 1.6 g ESI, and 140 mL water). The leather was hung on a stirring bar. When the bar rotates (50 rpm), the leather hits the glass wall, which to some extent simulates the action of drumming.

When ultrasound alone is applied to the preparation of fatliquors, a finer emulsion is produced. The mean particle size, measured by a Coulter Counter Particle Size Analyser after dilution, not *in situ*, is reduced by about 15%, as shown in Figure 16. Use of this sonicated fatliquor emulsion for a conventional fatliquoring process can increase the final fat content remarkably (see Table 1). The fat contents were also analyzed individually for the three splits or strata (0.6–0.7 mm) across the cross section from grain to flesh side, i.e., the grain, middle, and flesh split, to examine the fat distribution. The results in Table 1 clearly reveal that the increase in fat content is most significant in the middle split. This indicates that the fat penetration and distribution are improved by using ultrasound-treated fatliquor emulsions, apparently due to reduced particle size.

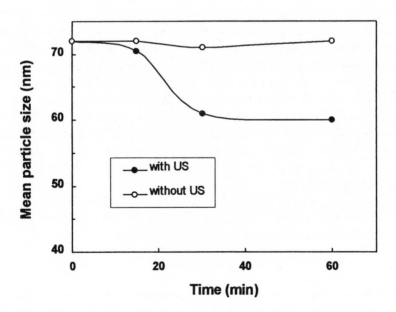

Figure 16. Effect of ultrasound on the mean particle size of a fatliquor emulsion (Remsynol ESI) at 40 °C.

Table 1. Relative Increase of Fat Contents in Different Splits of Bovine Wet-Blue Leather after Using a Presonicated Fatliquor Emulsion Compared with a Conventionally Prepared Fatliquor Emulsion[a]

Time	Flesh Split	Middle Split	Grain Split	Overall
30 min	3%	86%	53%	36%
60 min	1%	44%	29%	36%

Note: [a]The fatliquoring process was carried out at 40 °C with 8% offer of fatliquor (Remsynol ESI).

When ultrasound was directly applied to the fatliquoring process itself, some very interesting results were obtained. Table 2 shows that sonication alone is less effective than mechanical agitation in enhancing the fatliquoring process, but a combination of ultrasound and mechanical agitation can produce a synergistic effect. The increase in fat content as well as the improvement in the fat penetration and distribution can be clearly seen in Figure 17, where the fat has been stained with a dye (Sudan IV). The highest fat content is achieved when the ultrasound is used in the last 30 min of a 60-min fatliquoring process, increasing the fat content by 40% at 40 °C and 22% at 60 °C. In contrast to dyeing, for fatliquoring an initial ultrasonic treatment is less effective than a later treatment.

Kinetic studies also revealed that for a fatliquoring process at 60 °C, a 40-min continuously sonicated process is equivalent to a 60-min conventional one in terms of leather fat contents achieved. Thus, ultrasound can shorten the processing time by about 30%. Analysis of different splits shows, again, that the fat content increase is most significant in the middle split (Table 3).

It is, however, interesting to note that when the results in Tables 1 and 3 are compared, it is seen that the use of a presonicated fatliquor emulsion in a conventional process gives a larger overall fat content increase (36 versus 29%) as well as a larger fat content increase in the middle split (44 versus 33%) than the continuously sonicated fatliquoring process using a conventionally prepared fatliquor emulsion. A larger increase in the fat content of the middle split would indicate a

Table 2. Overall Fat Contents (% Dry Weight of Leather) after 60 min Fatliquoring with 8% Offer of Fatliquor (Remsynol ESI) for Bovine Wet-Blue Leather (Error Bar ±0.5%)

	40 °C	60 °C
Mechanical agitation	6.1	8.6
Sonication	4.4	5.6
Sonication and mechanical agitation (60 min)	7.8	10.3
Sonication in the first 30 min	7.4	9.0
Sonication in the last 30 min	8.5	10.5

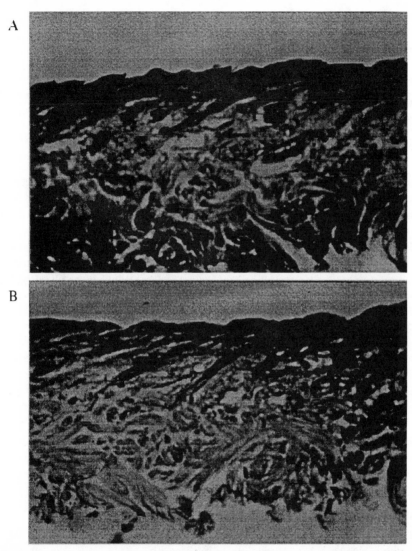

Figure 17. Micrographs showing the penetration of fatliquor into bovine wet-blue leather after a 60-min fatliquoring process at 40 °C, (top) with continuous sonication and mechanical agitation and (bottom) with mechanical agitation only. The fat was stained with Sudan IV dye.

better penetration. This finding suggests that the influence of particle size on the penetration of the fatliquor is more important than the direct effect of ultrasound. This argument is further supported by the observation that there is a difference between the fat content increases of the grain split from leather subjected to a

Table 3. Relative Increase of Fat Contents in Different Splits of Bovine Wet-Blue Leather after a Continuous Ultrasonic Fatliquoring Process, Compared with a Conventional Fatliquoring Process, at 40 °C with 8% Offer of Fatliquor (Remsynol ESI)[a]

Time	Flesh Split	Middle Split	Grain Split	Overall
30 min	36%	59%	5%	29%
60 min	13%	33%	–7%	28%

Note: [a]The fatliquor emulsion was not presonicated.

continuously sonicated fatliquoring process and the grain split from leather subjected to a conventional fatliquoring process using a presonicated fatliquor emulsion, i.e., a finer emulsion (see Tables 1 and 3). The grain layer consists of fine fibers and has a more compact structure than corium layer (mid or flesh split). Therefore, the use of a finer emulsion of smaller particle size would obviously tend to increase the fat penetration or diffusion. This is exactly what was observed in the conventional fatliquoring process using a presonicated fatliquor emulsion. But ultrasound per se does not appear to have a great influence on the fat penetration into the grain layer, at least at the power level of 1.3 W cm^{-2}. However, for the flesh split, the opposite effect is observed: The ultrasonic fatliquoring process with a conventional fatliquor emulsion leads to greater fat penetration than the conventional fatliquoring process with a presonicated fatliquor emulsion. Since the flesh side has coarser fibers and a looser structure than the grain does, it appears that ultrasound is more effective in enhancing the fat penetration when the particle size is relatively small compared with the pathway. This observation is, however, not fully understood. It would of course be interesting to further investigate an ultrasonic fatliquoring process with a presonicated fatliquor emulsion.

3.4 Tanning

Tanning was perhaps the most intensively investigated process in the 1950s and 1960s [5–8,10,12–15,19,20]. Early investigations have shown that ultrasound is not very useful for chrome tanning, but very effective in accelerating the vegetable tanning process. Vegetable tanning is the most time-consuming process in leather manufacturing. Modern vegetable tanning, especially for sole leather, is still conducted in pits, the traditional form of vegetable tanning, although the time needed has been reduced to a few weeks from more than 1 year at around the turn of this century. Vegetable tanning can also be conducted in a rotating drum, but the hide needs to be pretanned with syntans (synthetic analogues of tannins). This is to reduce the affinity of hides to vegetable tannin materials so as to facilitate the penetration.

Recent work by Xie et al. [34,35] on the effect of ultrasound (38 kHz, 1.3 W cm^{-2}) on vegetable (mimosa) tanning confirmed the powerful effect of ultrasound on speeding up the process. This can be seen in Figure 18, which shows the effect of ultrasound on the kinetics of mimosa tanning of bovine hides. The experiment was conducted in a glass vessel at 25 °C under a static condition similar to pits, and the ultrasound was only applied in the first 2 h. Typically, 60 g hide, 18 g mimosa tanning reagent (Hodgson Chemicals), and 360 mL water were used in the experiment. It can be seen that for a 2-h ultrasonic tanning process, the tannin content in leather has already reached the same level as for a 5-h conventional process. Ultrasound can thus shorten the process time by about 60%. The degree of penetration is increased by about 40%. The mean particle size of mimosa liquor was also found to be reduced from about 900 nm to 500 nm.

Current tanning technology is dominated by chrome tannage. The effect of ultrasound on this tannage is, unfortunately, not very great as has been shown by the early investigations [6,7]. Recent results [34,35] also showed that, as far as the rate of penetration of chromium complexes into the leather is concerned, the increase is on the order of 10%. The exhaustion of chrome liquor is also increased by a similar order; practically this would be more beneficial than the increase of penetration rate because it would help to reduce the effluent discharge. The amount of chrome leaching is also reduced by up to 20%, after using ultrasound, which would help to reduce the environmental impact when leather waste is used in landfill. It must be pointed out here, however, that although chromium(III) is perceived as an environmental hazard, it is actually nonhazardous and safe, unlike chromium(VI), which is very toxic. Perhaps the most important effect of ultrasound

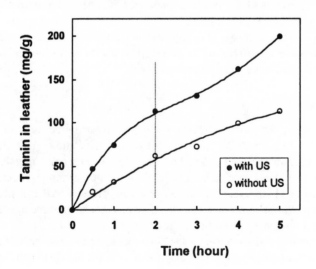

Figure 18. Effect of ultrasound on the vegetable tanning with 30% mimosa and 600% float at 25 °C for the first 3 h and 35 °C for the remaining time, without mechanical agitation; ultrasound being applied only in the first 2 h of the whole process.

on chrome tanning is that the shrinkage temperature is increased by 3–5 °C, compared with the same level of chrome content in conventionally tanned leather. This is probably due to a more even distribution of chromium in the leather, as shown in Figure 19. Thus, despite the negative results of early work, the use of ultrasound in chrome tanning may actually beneficially influence the quality of the leather produced and reduce the environmental costs. This is an area that needs further investigation.

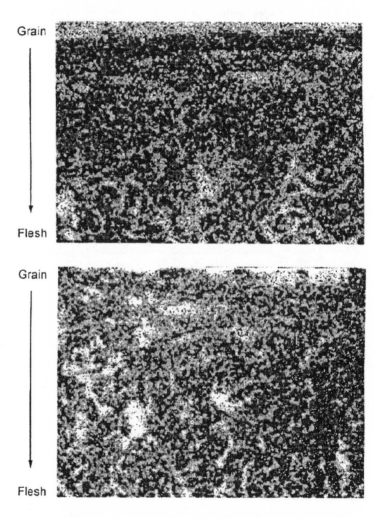

Figure 19. Chromium maps obtained by SEM/X-ray microanalysis, (top) with and (bottom) without ultrasound. The darker the shade is, the higher the local chromium concentration.

Mäntysalo et al. [32] recently reported interesting work on the application of power ultrasound to chrome tanning based on the use of a roller assembly as described by Kessler [42]. Kessler's process was one of the many attempts to develop a continuous flow production line (see, e.g., Heidemann [38]), in contrast to the currently dominating batch production process based on drums. The principle of this process is to use rollers to compress hides as they move through a tanning liquor. In the subsequent relaxation after compression, the hides suck in the liquor like a sponge. Thus, the rate of penetration is increased. Mäntysalo et al. found that irradiation of the hides with ultrasound (30 kHz, 600 W), immediately after a single rolling, can significantly improve the penetration and chrome uptake. Penetration can be completed in a few minutes. The resulting leather had a shrinkage temperature above 100 °C, which is difficult to achieve by a consecutive multirolling process. Thus, the single rolling combined with ultrasonic irradiation is a much better process than a consecutive multirolling process. This process should be equally applicable to other leather processes such as dyeing.

4. CONCLUDING REMARKS

It has been demonstrated by early and recent investigations that power ultrasound has great potential in leather technology. One of the main advantages of ultrasound lies in its general applicability to many leather manufacturing processes such as soaking, unhairing and liming, tanning, fatliquoring, dyeing, and so on. It has also been shown by recent work that the benefits of ultrasound are not only an accelerated process, but also savings in the environmental costs, amounts of chemical used, and improvements in product quality. In addition, there exist new dimensions for the application of ultrasound to the leather industry, such as effluent treatment.

The first sizable application of power ultrasound in the leather industry will probably be in the vegetable tanning pits and in the preparation of fatliquor emulsions of smaller particle size. These would seem to be relatively straightforward applications. However, the very large majority of leather in use today is processed in rotating drums. The scaleup of the laboratory-based procedures we have described in this review to full-size industrial drums is a difficult task and requires a considerable investment in research and development. However, the potential rewards of improved process times and savings in energy, chemical, and environmental costs we believe will make such an investment well worthwhile.

REFERENCES

[1] Sharphouse, J. H. *Leather Technician's Handbook*. Leather Producers' Association, London, 1971.
[2] Covington, A. D., and Alexander, K. T. W. *J. Am. Leather Chem. Assoc.*, 88 (1993) 241.
[3] Stanley, A. BLC The Leather Technology Centre, Northampton, U.K., 1993.
[4] Covington, A. D. *Chem. Soc. Rev.*, (1997) 111.
[5] Zapf, F. German Patent 840746 (1949).

[6] Ernst, R. L., and Gutmann, F. *J. Soc. Leather Technol. Chem.*, 34 (1950) 454.
[7] Gutmann, F. *J. Br. Inst. Radio Eng.*, (1955) 357.
[8] Fridman, V. M., Zaides, A. L., Mikhailov, A. N., Dolgopolov, N. N., and Karavaev, N. M. *Dokl. Akad. Nauk S.S.S.R.*, 92 (1955) 399.
[9] Realisations Ultrasoniques. British Patent 837521 (1956).
[10] Mieczyslaw, T. *Rev. Tech. Ind. Cuir*, 50 (1958) 261.
[11] Herfeld, H. *Gerbereiwiss Praxis*, 30 (1978) 163.
[12] Tielborger, H. German Patent 902169 (1954).
[13] Eisenegger, F. Swiss Patent 293120 (1953).
[14] Aksel'band, A. M., Gri, M. G., and Nozenko, A. N. *Kozh. Obuvn. Promot.*, 3 (1961) 24.
[15] Witke, F. *Ost. Leder-Ztg.*, 7 (1952) 165.
[16] Karpman, M. J. *Kozh. Obuvn. Promst.*, 4 (1962) 34.
[17] Alexa, G., Marinescu, M., Matei, E., and Luca, E. *Rev. Tech. Ind. Cuir.*, 56 (1964) 73.
[18] Khr, B. *Khim. Ind.*, 38 (1966) 67.
[19] Wenzinger, A. *Rev. Tech. Ind. Cuir*, 51 (1959) 237.
[20] Masner, L. *Kozarstvi,* 10 (1960) 328.
[21] Gourlay, P. *Rev. Tech. Ind. Cuir*, 51 (1959) 240.
[22] Senilov, B. V., and Obukhov, A. D. USSR Patent 133160 (1960).
[23] Metelkin, A. I., and Suchkov, V. G. *Kozh. Obuvn. Promst.*, 3 (1961) 27.
[24] Kotlyarevskaya, K. B., Maier, E. A., and Kondratenko, B. P. *Kozh. Obuvn. Promst.*, 6 (1964) 27.
[25] Green, G. H., and Moss, J. A. *Rev. Tech. Ind. Cuir*, 50 (1958) 106.
[26] Simoncini, E., and Criscuolo, I. *Cuoio Pelli Mater. Concianti*, 29 (1953) 82.
[27] Potoschnig, F., and Liebscher, E. Austrian Patent 178581 (1954).
[28] Schantz, H. German Patent 918654 (1954).
[29] Xie, J. P., Ding, J. F., Mason, T. J., and Attenburrow, G. E. *XXII IULTCS Congress Proceedings* Part II, p. 60, Friedrichshafen, Germany, 1995.
[30] ALPA S.p.A. *World Leather*, November 1995, p. 54; Monfrini, L. *48th Annual Convention Proceedings of the Society of Leather Technologists and Chemists* (South Africa Section), Drakensberg, Kwa Zulu Natal, South Africa, 1996.
[31] Xie, J. P., Ding, J. F., Mason, T. J., and Attenburrow, G. E. 5th Meeting of European Society of Sonochemistry, Cambridge, U.K., 1996; 1996 Annual Meeting of the American Leather Chemists Association, Michigan.
[32] Mäntysalo, E., and Marjoniemi, M. 5th Meeting of European Society of Sonochemistry, Cambridge, U.K., 1996; Mäntysalo, E., Marjoniemi, M., and Kilpeläinen, M. *Ultrasonics Sonochem.*, 4 (1997) 141.
[33] Ding, J. F. *World Leather*, April 1997, p. 57.
[34] Xie, J. P., Ding, J. F., Mason, T. J., and Attenburrow, G. E. *IULTCS Centenary Congress Proceedings*, p. 585, London, 1997.
[35] Xie, J. P. Ph.D. Thesis, Nene College of Higher Education, Leicester University, U.K., 1998.
[36] Ashiq, F. M.Sc. Dissertation, Nene College of Higher Education, Leicester University, U.K., 1996.
[37] Zainescu, G. A., Bratulescu, V., Georgescu, L., and Barna, E. *IULTCS Centenary Congress Proceedings*, p. 183, London, 1997.
[38] Heidemann, E. *Fundamentals of Leather Manufacture*, Eduard Roether KG, Darmstadt, 1993.
[39] Blair, T. *Leather Manuf.*, (1986) 18.
[40] Mason, T. J. *Practical Sonochemistry: A User's Guide to Applications in Chemistry and Chemical Engineering*, Ellis Horwood, Chichester, 1991.
[41] Vickerstaff, T. *The Physical Chemistry of Dyeing*, 2nd ed. Oliver and Boyd, London, 1954.
[42] Kessler, H. *Das Leder,* 25 (1974) 129.
[43] Thakore, K. A., Smith, C. B., and Clapp, T. G. *Am. Dyest. Rep.*, 79(10) (1990) 30.

SONIC ENERGY IN PROCESSING:
USE OF A LARGE-SCALE, LOW-FREQUENCY SONIC REACTOR

John P. Russell and Martin Smith

OUTLINE

Advances in Sonochemistry
Volume 5, pages 279–302.
Copyright © 1999 by JAI Press Inc.
All rights of reproduction in any form reserved.
ISBN: 0-7623-0331-X

1. INTRODUCTION

Many of the applications previously developed using sonic energy have not achieved their full potential due to lack of scaleup: To be fully practical in industry, these applications require the availability of new sonic equipment with more power, capable of greater throughput.

ARC has developed and patented this new equipment. This contribution describes the technology in detail and presents a range of applications that have been tested and are of interest to industry.

History of Development

The ARC Sonicator was conceived in response to work done in the early 1960s on vibratory milling. This work indicated a tremendous potential for energy-efficient grinding if a relatively high frequency (> 60 Hz), high-amplitude (up to 6 mm) vibratory ball mill could be built and operated reliably. ARC Sonics Inc., of Vancouver, Canada, has developed a reliable vibratory system that is capable of meeting these process requirements at the power levels required by industry. The development of the Sonicator was in response to a need to provide large-scale milling capabilities not possible with existing sonic equipment. The first piece of equipment built was the SonoGrinder (completed in 1991) for use in ore grinding.

ARC has three configurations that it currently uses for client test work and to evaluate sonic processes. The units are identified by their nominal power draw in kilowatts and are described in Table 1.

Worldwide equipment patents have been filed by ARC and issued in Europe, the United States, Canada, and many other countries, and are pending elsewhere. The technology is patented in two categories, as a Sonic Generator and as a Method and Apparatus for Grinding Material. Further patent applications are being investigated and will be filed in due course, covering second-generation equipment and refinements. A first process patent has been granted in the United States and other

Table 1. ARC Sonic Units Described

Nominal Power Draw (kW)	Mechanical Frequency (Hz)	Design Orientation	Masses (Tonnes)
5	300	Vertical bar as probe	< 1
20	440	Horizontal bar with two chambers	~ 2
75	100	Horizontal bar with two chambers	13.5

patentable processes are currently under review or are being actively investigated [1].

Various projects have been undertaken in the last 5 years either by the company or with industrial partners. Some are described below with client names omitted, for reasons of client confidentiality or possible patent filings.

2. THE TECHNOLOGY EXPLAINED

2.1 The Technology

2.1.1 Acoustic Energy Generation

The Sonicators (of which the SonoGrinder is a member) are a family of unique oscillating machines capable of generating high power at low frequencies and, when immersed in a liquid or slurry, creating intense cavitation and high-energy pressure waves. At the heart of each machine is a large, solid bar that is made to vibrate at its resonant frequency. The operating frequencies range from 100 Hz (large steel bar) up to 500 Hz (smaller bar), with amplitudes of up to 6 mm peak to peak at the lower end of the frequency range. The vertical unit operates like a large probe immersed in a reaction medium (Figure 1). The more powerful units are of horizontal design with reaction chambers bolted to each end of the bar (Figure 2). The frequency and amplitude of vibration in the horizontal mode results in very high inertial forces (Figure 3).

2.1.2 Power Requirements and Output

ARC Sonicators employ a sophisticated electromagnetic driving mechanism which overcomes the problems associated with mechanical drives (Figure 4). This drive system uses a variable-frequency inverter power supply. The use of this type of power supply enables the sonicators to operate at the peak efficiency resonant condition. This system has proven to be very effective. The earliest prototype has performed in service for more than 3000 h with virtually no maintenance.

In principle, the electromagnetic drive system is extremely simple. The drive units are energized at the desired natural mechanical resonant frequency of the

5 kW unit

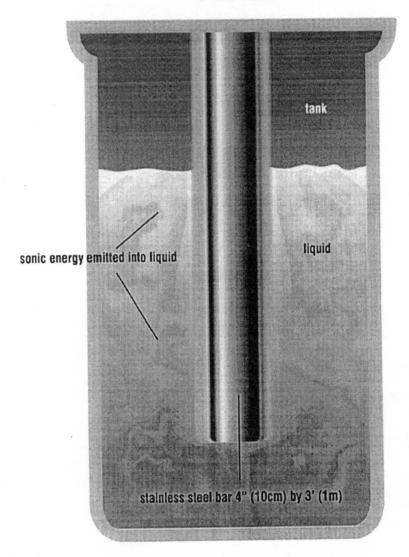

Figure 1. Vertical bar configuration.

massive steel bar by the variable-frequency, three-phase, AC inverter power supply. Each phase of the power supply energizes one of the three electromagnetic drive units. Each drive unit is spaced at 120° radially about the resonant member such that the magnetic force vector produced by the drive units rotates at a constant rate

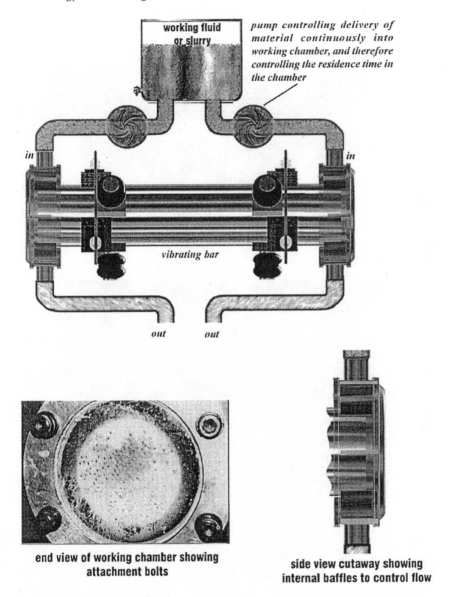

Figure 2. Bar with reaction chamber attached and continuous throughput.

about the longitudinal axis of the resonant system at the driven frequency. (Note: the force vector moves, the bar resonates but otherwise does not move.)

This causes a three-dimensional, rotational vibration of the resonant member which allows energy to propagate radially from the ends of the resonant bar in all

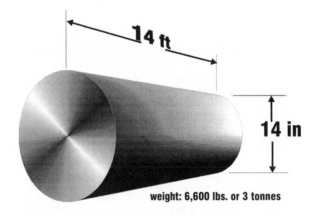

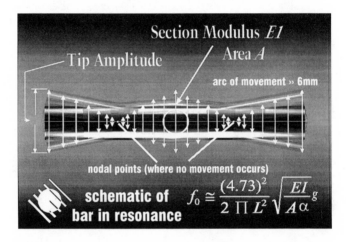

Figure 3. Diagram of bar on the large, 75-kW unit.

directions. Exciting the radiating bar at its natural resonant frequency results in a tremendously increased power transmission ability for a given excitation force; this is the key to the new technology (Figures 5 and 6). Standard industrial three-phase electrical power of 440 to 600 volts is required. Any reasonable three-phase industrial power source can be adapted as necessary.

2.1.3 Throughput

Every project to date has resulted in a different processing time or rate of throughput. The best indication of volume capability resulted from the grinding

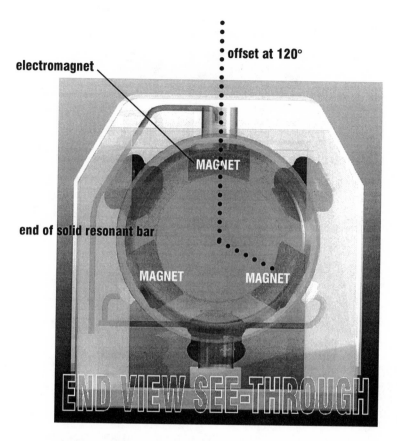

the electromagnets are timed to drive the bar at its resonant frequency

Figure 4. Configuration of driving electromagnets.

pilot test done in a mine mill: A hard ore at 500–300 µm was the feedstock, continuously processed as a slurry with retention time of 1.5 s, to give a product of 100 µm. After optimization, the volume treated was calculated at 100 tonnes dry weight per 24-h day. For physical processes like grinding, this figure is a guidepost; for sonochemical reactions, only detailed testing will give a useful figure.

2.1.4 Noise and Vibration

Since the equipment generates sonic energy, it is reasonable for users to inquire about its sound profile: When used for batch testing with the sound housing removed, the sound level is greater than 100 dB adjacent to the unit (ear protection is mandatory); when optimized for continuous processing and with the sound

dimensions: minimal length

low weight
stiff material like titanium

short bar, minimal mass, stiff material

result: high natural resonant frequency, possibly up to 1,500 Hertz

long, heavy bar of steel
result: lower frequency, down to below 100 Hertz

$$f_0 \cong \frac{(4.73)^2}{2\Pi L^2} \sqrt{\frac{E1}{A\alpha}} g$$

formula: frequency is a function of length, cross-sectional area, mass and modulus of elasticity

Figure 5. Natural resonant frequency.

housing properly in place, noise is approximately 60 dB. This can be easily improved for particular working situations as the sound generated is at a specific frequency.

The isolation of the bar by the use of pneumatic air bags minimizes vibrational losses to the frame. Accordingly, the units are placed unsecured on leveling pads: There is no movement of the unit as a result of vibration.

2.1.5 Footing Requirement

The footings or foundations employed today with the units are minimal—simple concrete or asphalt pads. The sonicators should be level in operation. Vibration is not a problem.

2.1.6 Portability

The equipment is transportable: The 5-kW unit weighs less than 1 tonne; the 20-kW, approximately 2 tonnes; the 75-kW, 13.5 tonnes. The 75-kW unit has been loaded onto a low-bed truck and transported 1600 km to Butte, Montana, from

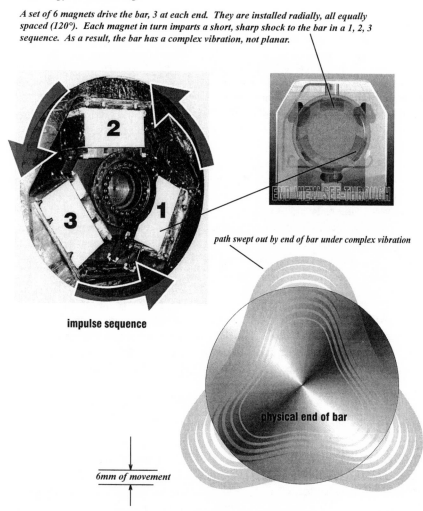

A set of 6 magnets drive the bar, 3 at each end. They are installed radially, all equally spaced (120°). Each magnet in turn imparts a short, sharp shock to the bar in a 1, 2, 3 sequence. As a result, the bar has a complex vibration, not planar.

impulse sequence

path swept out by end of bar under complex vibration

physical end of bar

6mm of movement

Figure 6. End vibration mode explained.

Vancouver, Canada. Here it was installed and tested, and was fully operational within 10 working days.

2.1.7 Reliability

The technology has developed to the point that the Sonicator is highly reliable. The 75-kW unit has been run for approximately 10,000 h to date with minimal repairs and maintenance.

2.1.8 Possible Future Developments

Throughout the development process the newer designs have run at higher frequencies resulting in significant increases in acoustic efficiency. The electromagnetic drive units have been improved to give higher power. Various reaction chambers have been tried and useful basic designs developed. The process of refining the design and developing application-specific modifications will continue.

New units are being designed with special features for particular applications. For example, an improved 20-kW design is completed, and further scaleup is under consideration up to 500 kW. Special configurations to work on larger volumes with lower power intensity are envisaged for use in wastewater systems, for instance. Because these are high-volume machines, considerable engineering will go into automated materials handling systems for both input and output—almost every application will require a special materials handling design.

2.2 The Sonic Phenomena

The ARC sonic technology relies on the efficient transfer of electrical energy to mechanical energy, sonic or vibrational, and then the extraction of that energy as useful work. When operated under resonant conditions, two phenomena have been demonstrated to take place: (1) high-impact pulverization or comminution and (2) cavitation.

2.2.1 High-Impact Pulverization or Comminution

A host medium is required to conduct the acoustic energy. Water is usually used. When grinding media are introduced into the reaction chamber, the sonic forces generated by the resonating bar are propagated off the walls of the chamber, through the fluid host medium, and so to the grinding media. The media collide with each other in a random manner with large impact forces, calculated as accelerations of up to 600 g. These extreme impacts have the effect of pulverizing any material located between the grinding media.

To prevent wear to the reaction chamber, a resilient liner is used. The liners have been found to wear very well and to protect the chamber walls for extended periods.

Since the resonant frequency of the system is in excess of 100 Hz, and the displacement of the grinding chamber is approximately 4–6 mm, the particle size reduction (comminution) of the material to be ground occurs extremely quickly, requiring, in most cases, residence times of only seconds (Figures 7 and 8).

Some materials with a range of particle sizes have a tendency to grind themselves, that is, the larger particles act as grinding media. This effect is known as *autogenous grinding*.

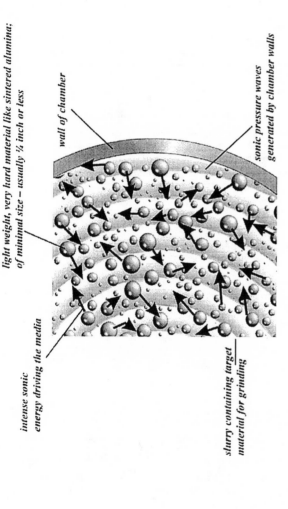

media
light weight, very hard material like sintered alumina;
of minimal size – usually ¼ inch or less

wall of chamber

sonic pressure waves
generated by chamber walls

intense sonic
energy driving the media

slurry containing target
material for grinding

Figure 7. Media movement resulting in fine grinding.

*result: movement of individual medium driven by the sonic pressure waves, random high energy collisions
and rebound collisions. The target material is impacted with massive force hundreds of thousands of times
per second.*

289

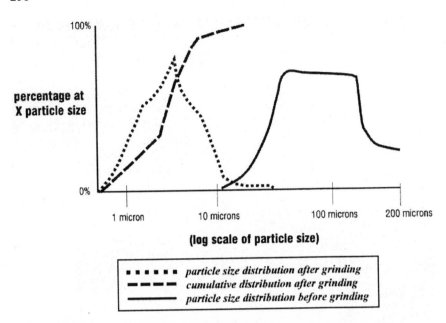

(based on extensive experience with Microtrack and Coulter Counter particle size analyses)

Figure 8. Grinding before/after particle distribution.

The grinding effect is enhanced by penetration of liquids into surface crevices in the particles. Sonic energy drives the liquids to penetrate into existing "micro cracks" and creates conditions in which the material will fracture and break.

2.2.2 Cavitation

The high sonic energy density in the reaction chamber causes cavitation, that is, microbubbles formed as a result of pressure waves. The process of cavity (bubble) collapse is so rapid that very high temperatures and pressures are created within the microbubble. Atchley and Crum [2] have reported that gas temperatures generated in the cavity are as high as 10,000 °C, giving rise to temperatures of 500 to 800 °C in the surrounding material, and temperatures as high as 5000 °C on the bubble surface have been suggested. Hunter and Bolt [3] state that shock waves, having a pressure difference of 4000 atm following a bubble collapse, have been demonstrated. Suslick has measured the cavitational bubble collapse of up to 106 events per ml under intense power. The effect of these thermal and pressure shocks on the material is erosion damage, crack propagation, and possibly mechanical failure.

2.3 Intellectual Property

The intellectual property described here is patented around the world. A listing of countries covered and of the state of the patents is given in the reference section.

3. PRACTICAL CONSIDERATIONS

3.1 Efficient Power Use

Approximately 60% of the electrical power draw is delivered to the reaction chambers in the form of useful acoustic energy. A minimal amount of power, measured at about 1–2%, is lost as vibration and sound. The larger power loss arises in conditioning the electrical energy and converting it to physical or vibrational energy. In order to condition the electrical power to drive the magnet assembly at the resonant frequency, its frequency is altered and the resulting waveform shaped. Various controls and operational sensors also require a small power draw.

3.2 Volume Processing

The key to successful volume processing is thorough initial testing to establish the optimum operating conditions. When these are known, the sonicator can be set up for continuous, maximum throughput.

It is possible to control the output results achieved by varying the power to the sonicator—this is not desirable. The most economic processing can be achieved when throughput is optimized and the unit is run at maximum power. In most applications a process control system is designed to feed back any change in output specifications, and to direct subtle changes in the operation to compensate for this variance. Control is achieved by changing input pumping rates so that the all-important residence time in the reaction chamber is altered; in this way a feed material can be given a specific residence time so that the end product continues to meet specification.

The most significant decision in relation to economic throughput is matching the size of the unit to the application so the unit can be run consistently at maximum load.

3.3 Chamber Design

Chambers are designed to contain a volume such that its mass will not overpower the resonance of the bar. The bar resonance, or Q feature, must be maintained without significant degradation, while being loaded to the maximum consistent with throughput. This is analyzed in the design stage where the following are considered: daily volume of product desired, its specific gravity when mixed with the host liquid to a suitable consistency, and the mass of the bar compared with the mass of the product in the chamber.

Chambers are designed with internal baffling to ensure that the material flows uniformly through the chamber, receiving consistent residence time; "tunneling" or dead spots are eliminated. The characteristics of the input material determine the design of baffles.

4. HETEROGENEOUS REACTIONS

4.1 Oxidative Destruction of Dyes Using Ozone and Low-Frequency Acoustic Energy

4.1.1 The Problem

Smith [4] has described the problem with dyes in this manner. Dyes used in the finishing of textiles are stable chemicals that have a strong color. The wastewater from textile plants can therefore be highly colored and difficult to treat by conventional biological and chemical processes. The environmental legislation on the color of wastewater in North America and throughout Europe is becoming more stringent. Regulations are in place, or are likely to come into place, in which consents to discharge contain specifications on the color of the wastewater from dyeing plants. The decolorization of waste dye solutions is therefore an area of concern in the textile dyeing industry, and others using dyes.

4.1.2 Prior Work

Physical treatments to remove color from wastewater, such as activated carbon and advanced oxidation methods using chemicals like chlorine or ozone, have been used, but not widely. Some investigative work has recently been carried out by Dr. Martin Smith at BC Research Inc. (a for-profit company located in Vancouver, British Columbia, Canada) on the oxidative destruction of dyes by low-frequency sonic energy and ozone using an ARC Sonics 20-kW unit, and Mason et al. [5] of the Sonochemistry Centre in Coventry University, England, have investigated the use of ultrasound with ozone.

4.1.3 Dye Destruction Test Program

This study was carried out to expand on the earlier work done at BC Research Inc. (using the ARC equipment) in order to determine whether the use of low-frequency sonic energy and ozone could successfully decolorize a range of textile dyes and enable them to be treated further by biological means. Other objectives included the investigation of operational parameters, particularly those that could be incorporated in the design of a continuous flow system, the commissioning of a continuous flow system, and comparison of the low-frequency sonicator with ultrasound probes. Where possible in this work it was the intention to apply methods used in the study of ultrasonic irradiation to low-frequency sonication.

An ARC Sonics 20-kW, 430-Hz sonicator fitted with a 3.2-liter reactor and supplied with ozone from a laboratory-scale generator was used for most of the investigations. Some work was also carried out on a laboratory scale at Coventry University using ultrasonic probes. Seven dyes were chosen for examination and were representative of those used by the textile industry. They were obtained from an international dye manufacturer trading in Canada and from a major textile company in England.

4.1.4 Results

The study demonstrated that water-soluble textile dyes can be destroyed and decolorized by low-frequency sonication combined with ozonation. Basic, acid, direct, and even some disperse dyes, which tend to have low solubility, were all rapidly decolorized. For most of the dyes the toxicity of the decolorized dye to microorganisms was the same as, or less than, the dye solution. This would make them suitable for further treatment on site or for immediate discharge to a publicly operated treatment plant. The amount of ozone required to achieve decolorization ranged from 1.1 to 1.9 kg of ozone per kg of dye. It is suggested that 2 kg of ozone per kg of dye be used in estimating costs and in the design of equipment. The rate of dye destruction was about 20 times faster than a simple mechanically stirred gas-to-liquid contactor. The rate is sufficiently great for short residence times of less than 60 s to be possible in a continuous flow process with industrial ozonizers and for dye concentrations within the range typical of wastewater from textile dyeing plants.

A continuous flow system for the sonicator reactor was commissioned and was used to show that the rate of dye destruction increased with increasing acoustic power input. In order to carry out this investigation it was necessary to develop a method of determining the acoustic power input into the reactor. The availability of such a method has allowed a meaningful comparison to be made with ultrasonic treatment of the dyes and will enable comparisons to be made with the performance of other low-frequency sonicators in the future.

4.1.5 Some Differences between Low and High Frequencies

Comparison of the ultrasonic treatment of dyes with that obtained at audible frequencies has indicated that low-frequency sonication does not appear to produce absolutely the same type of sonochemical reaction for the decomposition of water as ultrasound. This is perhaps not unexpected if we extrapolate back from the high generation of radical species at 1 MHz through the lower generation at 20 kHz and then down to audible. However, this does not mean that there are no sonochemical reactions brought about by low-frequency sonication. It seems likely that the high rates of dye destruction by sonication and ozonation may well involve the sonochemically aided decomposition of ozone.

4.2 Soil Remediation

The project work of Warren et al. [6] demonstrated the successful application of low-frequency acoustic energy to the cleanup of contaminated soils. As part of this project, partially funded by the Development and Demonstration of Site Remediation Technologies (DESRT) program sponsored by Environment Canada, ARC's reactors were used to expose hazardous organic contaminants trapped in soil, to the oxidizing agents hydrogen peroxide and ozone. The success of this project was a result of the sonicator's ability to disperse agglomerated soils, such as clays, to form near-perfect slurries of suspended soil particles.

Oxidation occurs in this process through the formation of hydroxyl radicals from intense sonic energy and chemical reactions with oxidation agents such as hydrogen peroxide and ozone. The hydroxyl radicals are powerful oxidizers that vigorously attack the chemical bonds in most organic compounds. Theoretically, the products that arise from this destruction technique are water and carbon dioxide, as well as hydrochloric acid in the case of chlorinated organic compounds.

Phase I of this project consisted of spiking clay with a known level of a particular contaminant and then subjecting the clay to a given set of treatment parameters. The four organic contaminants used were perchloroethylene (PCE), xylene, phenol, and pentachlorophenol (PCP). Testing revealed that the combination of ozone addition and sonic dispersion provided the best destruction rates. For PCE, Industrial Soil Remediation Criteria were achieved, while for xylene, the more stringent Residential Criteria were achieved.

Employing sonic agitation for dispersing the clay, using ozone as the oxidant, and starting with high PCP concentration, i.e., greater than 1000 ppm, destruction appeared to be inhibited when the concentration fell to the 100 ppm range. It was suspected that this occurred as a result of the formation of intermediates, which then reacted with the available oxidant therefore acting as scavengers. While the initial PCP remediation came close to achieving criteria levels, later work has overcome the limitations encountered. At lower initial concentration, e.g., 100 to 200 ppm, rapid destruction continued down to below 3 ppm.

In Phase II, some of the work on spiked soils, using ozone as the oxidant, was validated by conducting tests on clay or silty soils contaminated by actual spills of the following: the wood-preserving chemicals PCP and tetrachlorophenol (TCP), and the organic solvents benzene, toluene, ethylbenzene, and xylene, which commonly occur together and are referred to by the acronym BTEX.

The PCP destruction rates were comparable to the rates found in Phase I for clay soil contaminated to a comparable level. It is noted that the work done in Phase I was actually done on water-soluble NaPCP (in order to achieve full solvolysis), while the PCP from the spill site was likely to be dissolved in an oil, such as diesel, and would be water insoluble. The destruction rate for the TCP was similar to the PCP destruction rate. BTEX-contaminated soils were effectively treated, although the most volatile constituents, benzene and toluene, appeared to evaporate rapidly once the soil was exposed to air.

4.3 Treatment of Groundwater Containing an Organochlorine Contaminant

The use of sonic energy in the treatment of contaminated groundwater is exactly the same as in the treatment of slurried soils. In the following project, the client was a chemical company with a spill that had spread, over the years, into a groundwater plume. The testing was approached in a series of steps as described below.

4.3.1 *Four Steps to Testing*

1. Procurement of samples: ARC arranged for ten 4-liter water samples to be delivered to the test facility. Multiple samples were obtained and analyzed in order to ensure that the sample selected was representative of the situation.
2. Analysis of sample before treatment: Duplicate samples were analyzed for the organochlorine contaminant in order to establish the starting concentration. If the concentration of the sample was found to be significantly different from expected, resampling was done.
3. A series of batch experiments: The groundwater samples were processed using low-frequency sonic energy and the following variables were controlled:
   - residence time (e.g., separate tests with residence times of 1, 5, and 10 min)
   - type of oxidant (ozone and peroxide) or other reagent
   - reagent concentration

   The work plan allowed for a selection of scoping experiments designed to cover the range of processing possibilities. It was necessary to ensure that the data collected permitted sufficient information to indicate where the results are optimal—this leads directly to the operational conditions that will have to be implemented in the field.

   Samples were taken before, during, and at the end of each run for analysis. The sample temperature was monitored during each run. The efficacy of the process was determined by monitoring contaminant concentration using GCECD (gas chromatography).
4. Evaluation of results and report: All results were tabulated and interpreted immediately during the course of the project. A report was provided describing the experimental conditions and results. The report also contained a proposed program for a pump and treat process to be implemented in the field.

4.3.2 Results Achieved

The results were excellent, namely, a reduction of the organochlorine contaminant from 1 ppm to below conventional detection levels. Field implementation is now being considered by the client.

4.4 Advanced Oxidation

The destruction of PCP by ozone in a low-frequency (approximately 430 Hz) sonic reactor was studied with 100 ppm PCP aqueous solution. The experimental work emphasized the effect of sonication on ozone dissolution and decomposition and PCP destruction. In comparison with two typical ozone mixing methods, mechanical mixing of ozone and ozone bubbling alone, the ozone dissolution rate and PCP destruction rate were significantly enhanced by sonication. This enhancement is most likely attributed to the sonochemical effects that result from sonication.

Two different initial pH levels (9.4 and 12.0) were compared with respect to PCP destruction. It was found that a basic condition should be maintained for complete destruction of PCP by ozone because PCP becomes insoluble in an acidic condition. At high-pH (12.0) ozonation, a first-order reaction rate constant of 2.23 min L for PCP degradation with sonication was obtained versus 0.14 min L with absence of sonication (ozone bubbling alone) and 0.6 min L with mechanical stirring. Therefore, the ozone dosage requirement for PCP destruction with sonication is largely reduced because much shorter reaction time is needed.

For comparison with ultrasound, a number of experiments on sonolysis of water and PCP solution were conducted. Unlike ultrasound, low-frequency sonolysis of water did not create H_2O_2 (ultrasound will degrade water into various energetic components, and will create measurable hydrogen peroxide) nor was PCP degraded by sonolysis alone.

5. PHYSICAL REACTIONS

5.1 Fine Grinding and Deagglomeration

Fine powders, whether wet or dry, have a tendency in transportation or in storage to agglomerate, i.e., to come together electrostatically. The best use of these materials usually depends on these particles being separate to achieve their greatest reactivity with, for example, chemicals. Intense sonic energy is a way of deagglomerating powders into liquid suspensions.

ARC has done this successfully with pigments, chemicals, industrial minerals, and clays into liquids. While many industrial applications remain to be explored, ARC has proven the technology's ability to fine grind and deagglomerate materials.

In the area of fine grinding, the sonicator was used at the Golden Sunlight Mine, a mine in Butte, Montana, where it ground 100 tonnes per day dry weight of ore, from 300–500 μm down to 100 μm.

It has also ground sand from 100 μm down to 1 μm or less, suitable for use as silica fume. An electron micrograph of the results is shown in Figure 9.

It has also been used to grind limestone and talc from 250 μm down to 5 μm, which would enable the use of these materials as paper fillers; and the grinding of sodium metal in oil to create a dispersion.

The sonicator is capable of achieving these results because of the violent movement of the reaction chambers, at a rate of 104 to 430 Hz, which creates acoustic shock waves and cavitation where fine media act like an attrition mill. Impact forces up to 600*g* result in fast, fine grinding.

In the area of deagglomeration, the finest material dealt with was initially agglomerated at 30–50 μm. When complete, the mean was 0.88 μm and the median was 0.68 μm.

5 μ

size comparison bar is for 5 microns

Figure 9. Silica sand after sonic grinding. Bar = 5 μm.

The DESRT [6] report describes virtually perfect slurrying of clays with platelets of approximately 7 μm.

The explanation for deagglomeration is similar to grinding, but without the use of grinding media. The shock waves create extreme shearing forces that break up agglomerated material or destructure "tight" soils. Cavitation acts like pinpoint explosions within this slurry.

A number of specific examples can be quoted for the application of grinding and deagglomeration. One material when received was agglomerated to 30–50 μm; after acoustic treatment, the results showed a mean particle size of 0.88 μm and a median of 0.68 μm. This was done as an unoptimized batch test. Continuous production in an optimized system will give commercial volumes at the required standard.

Calcinated aluminum powders were ground in water to 90% less than 9 μm and 10% less than 2.3 μm. Based on unoptimized batch tests using the 75-kW unit, we estimated a throughput capacity of 3 tonnes per day.

In the case of paper fillers from calcium carbonate and talc, unoptimized batch processing gave 30% less than 5 μm. Continuous production in an optimized system in combination with classification is expected to result in commercial volumes meeting the industry specification.

A major company conducted tests and concluded that the ARC technology would be economic in grinding anthraquinone to the industry standard of 40 μm.

5.2 High-Shear Mixing and Emulsifying

As can be seen from the fine grinding example described above, the effect of the sonic energy in the reaction chamber is to severely agitate the contents. When no grinding media are used, and when the contents are liquids or suspensions, the effect of powerful, complex pressure waves and cavitation is to quickly mix the contents and shear or tear agglomerated materials. Preliminary work by ARC has illustrated that the result with an oil–water mixture is to emulsify; with pigments in solvent, to mix completely. There is some evidence that long-chain molecules are sheared and shortened.

5.3 Cell Lysing in Sludge Treatment

Municipal sewage treatment is frequently handled today by means of activated sludges. These sludges contain engineered bacteria that are successful in attacking and destroying the contaminants for which they are engineered. In the process, however, they multiply. This then creates a problem for the operators: disposal of the excess sludge.

Intense, low-frequency sonic energy has been used on these bacterial sludges with useful results. The sonically induced high-shear mixing breaks up the mats or rafts of bacteria and tears away the bound water or gel that is their principal defense against chemical and heat processing, the current standard method of destruction.

When the cells are separated and their bound water defenses dissipated, they are vulnerable to further treatment leading to their disposal. ARC is working on designing and testing a complete system for this purpose.

6. OTHER APPLICATIONS

6.1 Examples of Other Applications

Table 2. Other Applications

Application	Of Interest to . . .
Degas water and other solvents	Municipal sewage treatment
	Industrial smell control
	As a step to saturation, below
Saturate water and other solvents with gases	Food processing
	Hospitals
	Replacement for some cleaning solvents
Emulsify	Paint—waterborne polymers
	Food industry
Defoam	Pulp and paper
	Industrial processes
Breaking of molecules by extreme high shear mixing	Paint and plastic—changing of viscosity in polymers
High shear mixing	Many industries
Dissolve and solubilize	Chemicals
	Pharmaceuticals
	Food and beverage
Precipitation and flocculation	Municipal sewage
	Industrial effluent treatments

6.2 Enhancement of Weak Catalysts

Catalysts are widely used in industry but can be a considerable expense. There is a wide class of weaker catalysts that tend not to be used because of their lesser effect, although many of these are relatively inexpensive by comparison. It would be of interest to industry if they could use these less expensive catalysts and achieve the same effects as at present. (This concept has not been fully explored by ARC.) Based on the work of Guisnet et al. [7], it is proposed that sonic energy will drive many of the weaker catalysts into the range necessary for reactivity.

Some examples are:

- Zirconia (ZrO_2) acts as a catalyst in conjunction with sonic energy, and the activity is enhanced with Cr and Mn ions. It will, for example, promote transfer hydrogenation.

- Alumina (Al_2O_3) can assist with catalysis in 1-butene isomerization.
- Zeolites and TiO_2 assist in oxidative reactions.
- K10 montmorillonites may be effective as catalysts in Diels–Alder reactions.

7. CURRENT MAJOR PROJECTS

7.1 FBC Ash Conditioning

There are a number of different technologies used worldwide to generate electricity. The three primary methods are nuclear, water power, and fossil fuels. All of these raise environmental and economic issues: Nuclear power plants are becoming uneconomical to repair, and there is negative public reaction to disposal of their waste products; hydroelectric power is limited by availability of suitable water resources, and also presents environmental difficulties; burning of fossil fuel is still the most commonly employed method. The fuel sources range from natural gas to coal to peat. Natural gas, while clean, can be expensive and is a limited energy resource. Coal of varying quality is readily available around the world. However, as lower-quality coal with higher sulfur is used, the environmental problems increase.

A solution to these problems associated with coal combustion is the fluidized bed combustion (FBC) method. "One of the most promising energy-conversion options available today because it combines high efficiency of combustion of low quality fuels with substantially lower emissions of sulfur and nitrogen oxides," says CANMET (Canadian Centre for Mineral and Energy Technology) in its technology publication 01097. FBC does have one significant shortcoming: The ash produced is highly toxic and presents disposal difficulties. The current method of treatment of the ash is to blend it with water and transport it to a secure landfill for permanent storage. This has disadvantages:

1. The ash is only 40% wetted.
2. When wetted, the ash chemically reacts and heats up in the landfill, and continues to react each time it rains or water is added.
3. Landfill liners are often damaged and highly toxic runoff leaks into the groundwater so the landfill remains dangerous to human and animal life for years.

Low-frequency sonication results in fully wetted ash enabling the chemical reaction to proceed to completion within hours, not years. Carbon dioxide, which is produced in the generation of power, can be recycled into the mixture giving a fully inert material, limestone. Up to 6–7% by weight of the flue gas carbon dioxide can be used in this way resulting in a valuable reduction in "greenhouse gas" production.

ARC Sonics has entered into a joint contract with Nova Scotia Power, a power and with CANMET to verify that the successful bench tests can be scaled up to a commercially viable process.

7.2 Sodium Dispersion Manufacturing

Polychlorinated biphenyls (PCBs) are examples of a family of man-made chemicals with similar molecular structures, namely, a carbon backbone and multiple chlorine atoms. PCBs were frequently used to prolong the life of oil used in electrical transformers. Similar chlorinated compounds include PCP and TCP, used in wood preservation, and PCE, a widely used solvent. These compounds are related to other solvents—benzene, ethylbenzene, and xylene—that together pose a significant environmental hazard—they are toxic and carcinogenic. Extremely resistant to breakdown by natural processes, they accumulate in animal and human flesh. All were in high-volume production before their hazards were fully understood. PCBs are no longer manufactured and safer alternatives for many of the others have been found.

The current problem is the destruction of these chemicals whether in storage or in the environment. High-temperature incineration has been used but can release partially degraded compounds that are themselves toxic and hazardous. A safer method is the chemical disassembly of chlorinated organic molecules by the use of finely divided sodium in oil (sodium dispersion). It reacts with the chlorine atoms giving salt (NaCl) and, in the case of PCB, a carbon-based sludge that can easily be removed from the treated transformer oils. It can similarly be used to treat other hazardous organochlorides.

The present method of producing sodium dispersion is expensive and time consuming. The ARC technology can quickly create fine dispersions at a cost well below the current market. The same technology can accelerate the reaction between PCBs (or organochlorides generally) and sodium dispersion.

ARC has entered into a license agreement with Powertech, a subsidiary of B.C. Hydro, one of the largest producers of electric power in North America who have already processed millions of liters of PCB-containing transformer oils. It is expected that Japan will provide the first market for this sodium dispersion and for sonically assisted PCB destruction.

8. FUTURE CONSIDERATIONS

Of all of the future possible applications of the sonic technology, the broad area of sonochemistry is the most impressive. ARC expects in time to find its equipment incorporated into many industrial processes in a number of roles:

- Physically impacting the process stream by grinding, mixing, shearing, etc.
- Extraction of active ingredients from raw materials

- Accelerating chemical reactions where they create bottlenecks in the overall process
- Creating or modifying a final product by emulsifying, homogenizing, or changing the final form of the output.

Obviously, there is much work to be done to explore and exploit all of the possibilities in conjunction with a wide range of industrial partners.

REFERENCES

[1] US patent 5,318,228 (1994).
[2] Atchley, A. A., and Crum, L. A. In Suslick, K. S. (ed.), *Ultrasound: Its Chemical, Physical, and Biological Effects*. VCH Publishers, New York, 1988, pp. 1–64.
[3] Hunter and Bolt. *Sonics. John Wiley & Sons, New York, 1955, pp. 225–232.*
[4] Smith, M. The oxidative destruction of dyes using ozone and low frequency sound. Unpublished report by ARC Sonics and The Centre of Excellence for Sonochemistry, Coventry University, 1997.
[5] Lindley, J. In Mason, T. J. (ed.), *Sonochemistry: The Uses of Ultrasound in Chemistry*. The Royal Society of Chemistry, Cambridge, 1990, pp. 102–109.
[6] Warren, D., Russell, J. P., and Bastien, C. Y. Integration of enhanced oxidation and sonic mixing for treatment of contaminated soil. Unpublished report by Environmental Technology Advancement Directorate, Environment Canada, 1995.
[7] Guisnet, M., Barbier, J., Barrault, J., Bouchoule, C., Duprez, D., Perot, G., and Montassier, C. *Heterogeneous Catalysis and Fine Chemicals III*. Elsevier, Amsterdam, 1993.

INDEX